MODELING FLIGHT

The Role of Dynamically Scaled
Free-Flight Models in Support
of NASA's Aerospace Programs

Joseph R. Chambers

Library of Congress Cataloging-in-Publication Data

Chambers, Joseph R.
  Modeling flight : the role of dynamically scaled free-flight models in support of
NASA's aerospace programs / by Joseph R. Chambers.
      p. cm.
  "NASA SP 2009-575."
  Includes bibliographical references and index.
  1. Airplanes--Models--United States--Testing. 2. Wind tunnel models--United
States--Testing. 3. Aeronautics--Research--United States--Methodology. 4. United
States. National Aeronautics and Space Administration. I. Title.

  TL567.M6C48 2010
  629.101'1--dc22

                                          2009048942

ISBN 978-0-16-084633-5

90000

9 780160 846335

For sale by the Superintendent of Documents, U.S. Government Printing Office
Internet: bookstore.gpo.gov Phone: toll free (866) 512-1800; DC area (202) 512-1800
Fax: (202) 512-2104 Mail: Stop IDCC, Washington, DC 20402-0001

ISBN 978-0-16-084633-5

TABLE OF CONTENTS

ACKNOWLEDGEMENTS

I sincerely thank the dozens of current and retired employees of the National Aeronautics and Space Administration (NASA) and its predecessor, the National Advisory Committee for Aeronautics (NACA) who shared their invaluable corporate memory and personal files in the preparation of this monograph. The challenge of attempting to chronicle and document over 80 years of research and development activities was extremely daunting and would have been impossible were it not for the assistance and encouragement of these individuals. Their direct participation in this project has ensured the preservation and archival documentation of some of the most remarkable test activities ever conducted at the NASA Langley Research Center, the NASA Ames Research Center, the NASA Dryden Flight Research Center, and the NASA Wallops Flight Facility. Special thanks go to the following individuals, who contributed material for this publication: Long P. Yip, Norman L. Crabill, Mark A. Croom, John Sain, Donald B. Owens, Kevin Cunningham, Sue B. Grafton, Dan D. Vicroy, and Joseph L. Johnson, Jr.

Special thanks goes to Anthony M. Springer of NASA Headquarters for providing the encouragement and mechanism for this undertaking and to Lynn Bondurant and Dorothy Watkins of Paragon TEC for coordinating the preparation and administration of the effort. Sue B. Grafton and Frederick J. Lallman provided access to many historical documents and photographs at Langley, and Lana L. Albaugh, Edward M. Schilling, and Jack Boyd contributed at Ames. Thanks also to Garland Gouger and the staff of the Langley technical library for assistance in accessing documents. Photo archivist Alicia Bagby also provided access to photographs in the Langley collection. Graphics specialist Gerald Lee Pollard of Langley made available his outstanding talent and materials.

Thanks also to Mike Fremaux, Ray Whipple (retired), and Don Riley (retired) of Langley; Chuck Cornelison of Ames; Christian Gelzer of Dryden; Greg Finley of NASA Headquarters for their superb editing efforts and suggestions; and Janine Wise of NASA Headquarters provided professional layout and organization.

INTRODUCTION

The state of the art in aeronautical engineering has been continually accelerated by the development of advanced analysis and design tools. Used in the early design stages for aircraft and spacecraft, these methods have provided a fundamental understanding of physical phenomena and enabled designers to predict and analyze critical characteristics of new vehicles, including the capability to control or modify unsatisfactory behavior. For example, the relatively recent emergence and routine use of extremely powerful digital computer hardware and software has had a major impact on design capabilities and procedures. Sophisticated new airflow measurement and visualization systems permit the analyst to conduct micro- and macro-studies of properties within flow fields on and off the surfaces of models in advanced wind tunnels. Trade studies of the most efficient geometrical shapes for aircraft can be conducted with blazing speed within a broad scope of integrated technical disciplines, and the use of sophisticated piloted simulators in the vehicle development process permits the most important segment of operations—the human pilot—to make early assessments of the acceptability of the vehicle for its intended mission. Knowledgeable applications of these tools of the trade dramatically reduce risk and redesign, and increase the marketability and safety of new aerospace vehicles.

Arguably, one of the more viable and valuable design tools since the advent of flight has been testing of subscale models. As used herein, the term "model" refers to a physical article used in experimental analyses of a larger full-scale vehicle. The reader is probably aware that many other forms of mathematical and computer-based models are also used in aerospace design; however, such topics are beyond the intended scope of this document. Model aircraft have always been a source of fascination, inspiration, and recreation for humans since the earliest days of flight. Within the scientific community, Leonardo da Vinci, George Cayley, and the Wright brothers are examples of early aviation pioneers who frequently used models during their scientific efforts to understand and develop flying machines. Progress in the technology associated with model testing in worldwide applications has firmly established model aircraft as a key element in new aerospace research and development programs. Models are now routinely used in many applications and roles, including aerodynamic data gathering in wind tunnel investigations for the analysis of full-scale aircraft designs, proof-of-concept demonstrators for radical aeronautical concepts, and problem-solving exercises for vehicles already in production. The most critical contributions of aerospace models are to provide confidence and risk reduction for new designs and to enhance the safety and efficiency of existing configurations.

NASA and its predecessor, the NACA, have been leading contributors to the technology of aerospace model testing for over 80 years. Virtually every technical discipline studied by the NACA and NASA for application to aerospace vehicles has used unique and specialized models, including the fields of aerodynamics, structures and materials, propulsion, and flight controls. These models have been used for thousands of investigations in a myriad of specialty areas as far ranging as aerodynamic and hydrodynamic drag reduction, high-lift systems, loads because of atmospheric gusts and landing impact, aircraft ditching, buffeting, propulsive efficiency, configuration integration, icing effects, aeroelasticity, and assessments of stability and control. Within these specialty areas, models can be further classified as "static" models or "dynamic" models.

Static wind tunnel models of aircraft and spacecraft are designed to extract high-quality, detailed aerodynamic data for analysis of their full-scale counterparts at specified flight conditions. Static tests are typically conducted with the model fixed to an internally mounted, force-measuring device known as an electri-

cal strain-gauge, which is in turn mounted to a sting support system. The orientation of the model and its support system relative to the wind tunnel flow is controlled by the investigator as required for specified aerodynamic test conditions. The focus of the static experiment is on obtaining detailed aerodynamic measurements for combinations of airspeed, model attitudes, and configuration variables such as wing-flap or control surface deflections. Extensive instrumentation is used to measure critical aerodynamic parameters, which may include forces and moments, static and dynamic pressures acting on the surfaces of the model, the state of the boundary layer, and visualization of flows on and off the surfaces

In this representative conventional static wind tunnel test, a 2.7-percent scale model of the Boeing 777 airplane is mounted to a sting support system for testing at transonic speeds in the National Transonic Facility at the NASA Langley Research Center. The model contains extensive instrumentation and an internally mounted electronic strain-gauge balance for measurements of aerodynamic data. In static testing, data are taken at specific fixed combinations of airspeed, model configuration, and model attitude as controlled by the test team.

Dynamic free-flight model tests of the X-29 advanced technology demonstrator were conducted in the NASA Langley Full-Scale Tunnel to determine the configuration's dynamic stability and control characteristics. The photograph shows the model flying at a high angle of attack while researchers assess its behavior. Note the large nose-down deflection of the canard surface, an indication of the inherent static instability of the configuration. Active controls using data from an angle-of-attack sensor on the nose boom provided artificial stability.

of the model. Design challenges for this class of model include providing sufficient structural strength, especially to accommodate the large loads encountered for high airspeeds, high angles of attack, or sideslip; protecting the model and its data sources from aerothermal loads; providing robust and reliable instrumentation within the relatively small airframe of the model; and designing the model components and test techniques to permit timely wind tunnel operations for acquisition of data with minimum "dead time" required for reconfiguration of the model or support system. An additional design issue is the required accuracy of model lines and deciding which airplane features need to be replicated for the model. The results of static tests are used in analyses or as inputs for subsequent investigations that may involve other testing techniques and additional methods. Note that static tests do not require dynamic motions of the model or free-flight capability.

Dynamic free-flight model investigations extend the objectives of conventional static tests to include the effects of vehicle motions. The primary objective of free-flight testing is to assess the inherent flight motions of a configuration and its response to control inputs. In accomplishing this task, the same models used for

Dynamic models have been used for over 50 years by the NACA and NASA to predict the spin entry and spin recovery behavior of highly maneuverable military aircraft. A model of the F-15 fighter is shown mounted on a launching rig attached to a helicopter in preparation for an unpowered "drop" test, during which the unpowered model will be controlled by ground-based pilots in deliberate attempts to promote spins. As a result of extensive correlation of results with subsequent full-scale aircraft results, NASA has validated its testing procedures.

free-flight tests can frequently be used in conventional static tests as well as unique wind tunnel test techniques to determine the effects of vehicle motions on critical aerodynamic parameters. The results of free-flight model tests can be readily observed by the motions of the model in response to forcing functions by the atmosphere or control deflections by pilots. Dynamic tests focus on critical characteristics such as stability, controllability, and safety during flight within and outside the conventional flight envelope. The most critical applications of free-flight models include evaluations of radical or unconventional designs for which no experience base exists, and the analysis of aircraft behavior for flight conditions that are not easily studied with static testing techniques or advanced computational methods because of complex aerodynamic phenomena that cannot yet be modeled. Examples of such conditions include flight at high angles of attack, stalls, and spins in which separated flows, nonlinear aerodynamic behavior, and large dynamic motions are typically encountered.

Even a cursory examination of the breadth of the NACA and NASA studies to date in static and dynamic model testing reveals an overwhelming wealth of experiences and technology far beyond the intended scope of this monograph. The objective of the material presented here is to provide the reader with an overview of some of the more interesting free-flight model testing techniques that have been developed and the role that the testing has played in fundamental and applied research, as well as in support of the development of some of the Nation's more important civil and military aerospace programs. The material also includes discussions of the development of the specialized facilities and equipment required for dynamic model tests.

These efforts originated over 80 years ago at the first NACA laboratory, known as the Langley Memorial Aeronautical Laboratory, at Hampton, VA. Most of the principles and concepts introduced by the NACA have been continually updated and are now fully incorporated within the testing capabilities of the NASA Langley Research Center. In addition to the developmental efforts and existing operational capability at Langley, this paper will also discuss dynamic free-flight model studies at the NASA Dryden Flight Research Center at Edwards Air Force Base, CA, the NASA Ames Research Center at Mountain View, CA, and the NASA Wallops Flight Facility at Wallops Island, VA. Within these NASA research organizations, years of experience using subscale free-flight models have been accumulated and correlated wherever possible with results of full-scale aircraft flight tests to establish the accuracy and limits of model testing.

Free-flying dynamically scaled models have evolved from simple unpowered silk-span-covered balsa replicas to highly sophisticated miniature aircraft with advanced materials, propulsion units, control systems, and instrumentation. This model of the X-48B blended wing-body configuration is typical of the current state of the art in dynamic model technology.

The material selected for review covers two major types of dynamic model tests referred to as dynamic free-flight tests and dynamic force tests. Free-flight tests involve testing techniques in which powered or unpowered models are flown without appreciable restraint in wind tunnels or at outdoor test ranges. The models may be uncontrolled, or controlled by remotely located human pilots. Dynamic free-flight models are designed and constructed in compliance with required scaling laws to ensure that dynamic motions are similar between the model and its full-scale counterpart. A brief review of the scaling procedures for dynamic models is included for background. Dynamic force tests are special wind tunnel test techniques developed by the NACA and NASA to measure the effects of vehicle motions on critical aerodynamic data for flight conditions during normal operations and situations beyond the normal flight envelope, such as spins and poststall gyrations. Several unique test apparatuses have been implemented to permit dynamic force tests in several wind tunnel facilities.

The organization of the material for each of the discussion topics includes a brief background on the objectives and applications of dynamic models in that area of interest, the unique results obtained for specific programs, and how such data are used in the research and development process.

The subject matter first reviews the special geometric- and mass-scaling requirements for free-flight dynamic models. Limitations of the free-flight techniques are also reviewed. The discussion then presents a review of the history of the development of free-flight techniques by the NACA and NASA, including discussions of the physical characteristics of key facilities, the models used, and the approach to testing. The major portion of the book focuses on the applications of the techniques to generic configurations for general research as well as to specific aircraft and spacecraft configurations. Important lessons learned are also reviewed. A brief review of the special dynamic force tests that usually accompany free-flight tests is provided, and the document concludes with the author's perspective on potential future opportunities and requirements in free-flight model testing.

The contents presented herein are intended for a wide range of interested readers, including industry teams within the civil and military aviation communities, high school and college students, designers and builders of homebuilt experimental aircraft, aviation enthusiasts, and historians. To provide the most benefit for this broad audience, the technical language used has been intentionally kept at an appropriate level, and equations and technical jargon have been minimized. Extensive references are provided for those interested in obtaining a deeper knowledge of subject matter presented in each major section.

During his career as a NASA researcher and manager, the author was honored to have made many presentations to the international aviation community and public on NASA's applications of dynamically scaled free-flight models in its aerospace research programs. During those occasions, the audience was always interested and intensely inquisitive as to the methods used to conduct meaningful model flight tests in research activities. On many occasions, the audience recommended that a publication on NASA's experiences with the free-flight models be prepared. It is sincerely hoped that the contents of this publication provides information that will be useful in clarifying this very valuable and historic technical capability.

CHAPTER 1:

BACKGROUND

Objectives

Dynamically scaled free-flying models are especially well suited for investigations of the flight dynamics of aircraft and spacecraft configurations. Flight dynamics focuses on the behavior of vehicles in flight and the causal factors that influence the motions exhibited in response to atmospheric disturbances or pilot inputs. The science includes aspects of applied aerodynamics, static and dynamic stability and control, flight control systems, and handling qualities. Applications of the free-flight test capability in flight dynamics have ranged from fundamental studies—such as investigations of the effects of wing sweep on dynamic stability and control—to detailed evaluations of the flight behavior of specific configurations—such as the spin resistance and spin recovery characteristics of highly maneuverable military aircraft. During these investigations, the test crew assesses the general controllability of the model and inherent problems such as uncontrollable oscillations or failure to recover from spins. The crewmembers also evaluate the effects of artificial stability augmentation systems, airframe modifications, and auxiliary devices, such as emergency spin recovery parachutes. Time histories of motion parameters—including model attitudes, angular rates, linear accelerations, angles of attack and sideslip, airspeed, and control positions—are recorded for analysis and correlation with full-scale results and other analysis methods. Aerodynamic data for the free-flight model can be extracted from the model flight results, and most models can be used in conventional static wind tunnel tests as well as specialized dynamic force tests. The results of free flight testing are extremely valuable during aerospace vehicle development programs because potentially undesirable or even catastrophic behavior can be identified at an early design stage, permitting modifications to be incorporated into the design for satisfactory characteristics. In addition, the data provide awareness and guidance to test and evaluation organizations of potential problems in preparation for and during subsequent flight tests of the full-scale vehicle. The payoff is especially high for unconventional or radical configurations with no existing experience base.

Predictions obtained from dynamic model tests are routinely correlated with results from aircraft flight-testing. In a classic NASA study of spinning characteristics of general-aviation configurations, tests were conducted with, right to left: a spin tunnel model, a radio-controlled model, and the full-scale aircraft. Langley test pilot Jim Patton, center, and researchers Jim Bowman, left, and Todd Burk pose with the test articles.

Role of Free-Flight Testing

Free-flight models are complementary to other tools used in aeronautical engineering. In the absence of adverse scale effects, the aerodynamic characteristics of the models have been found to agree very well with data obtained from other types of wind tunnel tests and theoretical analyses. By providing insight as to the impact of aerodynamics on vehicle dynamics, the free-flight results help build the necessary under-standing of critical aerodynamic parameters and the impact of modifications to resolve problems. The ability to conduct free-flight tests and aerodynamic measurements with the same model is a powerful advantage for the testing technique. When coupled with more sophisticated static wind tunnel tests, com-putational fluid dynamics methods, and piloted simulator technology, these tests are extremely informative. Finally, the very visual results of free-flight tests are impressive, whether they demonstrate to critics and naysayers that radical and unconventional designs can be flown or identify a critical flight problem for a new configuration.

As might be expected, conducting and interpreting the results of free-flying model tests requires expertise gained by continual feedback and correlation with fight results obtained with many full-scale aircraft. Model construction techniques, instrumentation, and control technologies are constantly being improved. Facili-ties and equipment must be updated, and radical unconventional aircraft configurations may require modi-fications to the facilities and testing techniques. NASA staff members have accumulated extensive exper-tise and experience with the technology of free-flight models, and as a result, the aerospace community seeks them out for consultation, advice, and cooperative studies. The physical facilities and human resources acquired and nurtured by the Agency in the science and art of free-flight testing has made extremely valuable contributions to civil and military programs.

Later sections of the document will provide overviews of the conception and development of free-flight testing techniques by NACA and NASA researchers as well as selected contributions made to specific aerospace programs. Before these topics are discussed, however, a review of the specific scaling proce-dures required for dynamic free-flight models will provide additional background and understanding of the model construction details and the testing procedure. In addition, observations regarding the constraints and limitations involved in free-flight model testing are presented.

Principles of Dynamic Scaling

To obtain valid results from free-flight model test programs, the model must be designed and constructed in a specific manner that will ensure that its motions are similar to motions that would be exhibited by the full-scale article. Dynamically scaled free-flight models are not only geometrically scaled replicas; they are specially designed to ensure motion similitude between the subscale model and the full-scale subject. When a geometrically similar model of an aircraft reacts to external forces and moves in such a manner that the relative positions of its components are geometrically similar to those of a full-scale airplane after a proportional period of time, the model and airplane are referred to as "dynamically similar," bringing about a condition known as dynamic similitude. When properly constructed and tested, the flight path and angular displacements of the model and vehicle will be geometrically identical, although the time required for selected motions of each will be different and require the application of mathematical factors for interpreta-tion of the results.

The requirements for scaling may be visualized with the help of the accompanying graphic, which shows the interaction between aerodynamic and inertial moments for an aircraft in a steady spin. The sketches indicate the physical mechanisms that determine the balance of moments that occur about the pitch axis

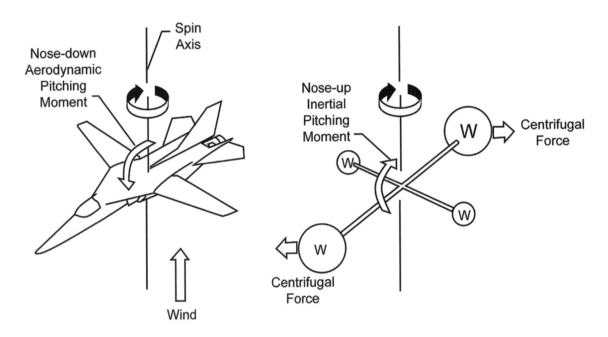

Sketch illustrating the balance of nose-down aerodynamic pitching moments and nose-up inertial pitching moments during a steady spin. The balance of these moments will largely determine the spin rate and angle of attack of the spin.

during the spin. The sketch on the left shows that, at spin attitudes, an airplane experiences very high angles of attack, which usually results in a nose-down aerodynamic pitching moment. This nose-down moment is balanced by the inertial pitching moment depicted in the sketch at the right. In the right-hand sketch, the mass distribution of the airplane is represented by a system of weights along the fuselage and wings. As the weights rotate about the spin axis, centrifugal forces acting perpendicular to the spin axis create a nose-up inertial pitching moment, attempting to align the fuselage perpendicular to the spin axis. For the steady developed spin, the nose-up inertial moment balances the nose-down aerodynamic moment.

It is readily apparent that the balance of aerodynamic and inertial parameters must be similar between the model and full-scale aircraft if the spin rate and angle of attack are to be dynamically reproduced. If, for example, the mass distribution of the model results in inertial nose-up loads that are too low compared with scaled airplane loads, the aerodynamics of the model will be grossly out of balance and possibly dominate the conditions of motion. On the other hand, if the aerodynamic moments are insufficient, the inertial moments will unrealistically determine the motions. Obviously, predictions of full-scale characteristics based on these conditions would be in error.

In addition to geometric scale requirements for the model (such as wingspan, length, and wing area), scale ratios of force, mass, and time must also be maintained between the model and full-scale article for dynamic similitude.[1] The fundamental theories and derivation of scaling factors for similitude for free-flight models are based on the broad science known as dimensional analysis. The derivations of dynamic-model scale factors from dimensional analysis were developed during the earliest days of flight and will not be repeated here. Rather, the following discussion provides an overview of the critical scaling laws that must be maintained for rigid (nonaeroelastic) free-flight models in incompressible and compressible flow fields.[2]

Rigid-Body Dynamic Models

The simplest type of dynamic free-flight model is known as the rigid-body model. As implied by the name, a rigid-body model does not attempt to simulate flexible structural properties such as aeroelastic bending modes or flutter properties of the full-scale article. However, even these simple models must be scaled according to mandatory relationships in each of the primary units of mass, length, and time to provide flight motions and test results directly applicable to the corresponding full-scale aircraft. Units of model length such as wingspan are, of course, scaled from geometric ratios between the model and full-scale vehicle, units of model mass (weight) are scaled from those of the full-scale vehicle on the basis of a parameter known as the relative density factor (measure of model mass density relative to an atmospheric sample), and model time is scaled on the basis of a parameter known as the Froude number (ratio of inertial to gravitational effects). From these relationships, other physical quantities such as linear velocity and angular velocity can be derived. For most applications, the model and full-scale aircraft are tested in the same gravitational field, and therefore linear accelerations are equal between model and full scale. To conduct meaningful tests using free-flight models, the scaling procedures must be followed during the construction, testing, and data analysis of a dynamic model. Simply scaling geometric dimensional characteristics without regard for other parameters can produce completely misleading results.

Incompressible Flow

Many of the free-flight dynamic model tests conducted by the NACA and NASA in research efforts have involved investigations of rigid models for conditions in which Mach number and compressibility were not major concerns. For these incompressible flow conditions, the required scale factors for dynamic models are given in the following table:

SCALE FACTOR	
Linear dimension	n
Relative density ($m/\rho l^3$)	1
Froude number (V^2/lg)	1
Angle of attack	1
Linear acceleration	1
Weight, mass	n^3/σ
Moment of inertia	n^5/σ
Linear velocity	$n^{1/2}$
Angular velocity	$1/n^{1/2}$
Time	$n^{1/2}$
Reynolds number (Vl/v)	$n^{1.5}v/v_0$

Scale factors for rigid dynamic models tested at sea level. Multiply full-scale values by the indicated scale factors to determine model values, where n is the ratio of model-to-full-scale dimensions, σ is the ratio of air density to that at sea level (ρ/ρ_0), and v is the value of kinematic viscosity.

Certain characteristics of rigid-body dynamic models are apparent upon examination of the factors given in the table. The required parameters of relative density and Froude number are maintained equal between model and aircraft (scale factor of 1), as are linear accelerations such as gravitational acceleration. The model and aircraft will exhibit similar flight behavior for the same angles of attack and dynamic motions. For example, following a disturbance at a specific angle of attack, the ensuing damped oscillations will be identical in terms of the number of cycles required to damp the motions for the model and aircraft.

As shown in the below table, however, other motion parameters vary markedly between model and aircraft. For example, for a 1/9-scale model (n=1/9), the linear velocities of the model (flight speeds) will only be 1/3 of those of the aircraft, but the angular velocities exhibited by the model in roll, pitch, and yaw will be 3 times faster than those of the airplane. Because the model's angular motions are so much faster than those of the airplane, the models may be difficult to control. Many dynamic model testing techniques developed by NASA meet this challenge by using more than one human pilot to share the piloting tasks.

Another important result of dynamic scaling is that large differences in the magnitude of the nondimensional aerodynamic parameter known as Reynolds number (ratio of inertial to viscous forces within the fluid medium) will occur. In other words, as a result of dynamic scaling, the model is tested at a much lower value of Reynolds number than that of the full-scale airplane for comparable flight conditions. A 1/9-scale dynamic model is typically tested at a value of Reynolds number that is only 1/27 that of the airplane for sea level conditions. As will be discussed, this deficiency in the magnitude of Reynolds number between model and aircraft can sometimes result in significant differences in aerodynamic behavior and lead to erroneous conclusions for certain aircraft configurations.

An examination of design specifications for dynamically scaled models of a typical high-performance military fighter and a representative general-aviation aircraft provides an appreciation for the impact of dynamic scaling on the physical properties of the models:

FIGHTER MODEL (N=1/9)	AIRPLANE	MODEL
Geometric scale	1.0	1/9
Wingspan	42.8 ft	4.8 ft
Weight (a)	55,000 lbs	75.5 lbs
Weight (b)	55,000 lbs	201.9 lbs
GENERAL-AVIATION MODEL (N=1/5)	AIRPLANE	MODEL
Geometric scale	1.0	1/5
Wingspan	36 ft	7.2 ft
Weight (c)	2,500 lbs	20 lbs
Weight (d)	2,500 lbs	27 lbs

Design results for models of representative rigid-body, general-aviation, and fighter models. Notes for weight data: (a) denotes full-scale airplane at sea level and model at sea level, (b) full-scale airplane at 30,000 ft and model at sea level, (c) full-scale airplane at sea level and model at sea level, and (d) full-scale airplane at 10,000 ft and model at sea level.

These data vividly demonstrate that dynamically scaled free-flight models weigh considerably more than would conventional hobbyist-type, radio-controlled models of the same geometric scale, and that the task of using a model at sea level to simulate the flight of a full-scale aircraft at altitude significantly increases the design weight (and therefore flight speed) of the model. The scaling relationships result in challenges for the dynamic-model designer, including ensuring adequate structural strength to withstand operational loads and crashes; providing adequate onboard space for instrumentation, recovery parachutes, and control actuators while staying within weight constraints; and matching flight speeds with those of wind tunnel facilities. Typically, a large number of preliminary trade studies are conducted to arrive at a feasible model scale for the altitude to be simulated in the flight tests. In some cases, the final analysis may conclude that the weight/payload/strength compatibility issues cannot be met for the type of testing under consideration.

In addition to addressing weight considerations, the designer must also meet scaling laws to simulate the mass distribution properties (moments of inertia) expected for the full-scale aircraft. Using data or estimates of the weight and center of gravity of various model components and equipment, calculations are initially made of mass moments of inertia about each of the three aircraft axes to determine inertias in pitch, roll, and yaw for the model. After the model is fabricated, experimental methods are used to measure the moments of inertia of the model, with all systems installed to ensure that the values obtained approach the values prescribed by the scaling requirements. The experimental methods used to measure inertias are based on observations of the times required for periodic oscillations of the model when forced by a mechanical spring apparatus, or pendulum-type motions of the model mounted to suspension lines. Using theoretical relationships for the motions and period of the oscillation, the inertias can be calculated. The inertia requirements and distribution of mass in the model can be a very severe design challenge, especially for powered models with heavy engines displaced along the wing.[3]

While the model satisfied geometrical, mass, and weight scaling constraints, the designer may also have to provide an appropriate propulsion system. Propulsion techniques used by NASA in its flying models have included unpowered gliders, conventional internal-combustion engines, tip-driven fans powered by compressed air, compressed-air thrust tubes and ejectors, turbojets, and electric-powered turbofans. The use of power systems and the degree of sophistication involved depends on the nature of the tests and the type of data desired. For example, free-flight investigations of spinning and spin recovery are conducted with unpowered models, whereas free-flight studies of powered-lift vertical take-off and landing (VTOL) aircraft require relatively complex simulation of the large effects of engine exhaust on the aerodynamic behavior of a model. Model propulsion units used to simulate current-day turbofan engines are especially complex.

Another challenge for the designer of a dynamic free-flight model is the selection of control actuators for aerodynamic control surfaces. The use of a proportional control system in which a deflection of the control stick results in a proportional deflection of the control surface may be suitable for precision control of larger models. However, because small, dynamically scaled models exhibit rapid angular motions, the use of rapid full-on, full-off "flicker" type controls may be required.

Compressible Flow

Critical aerodynamic parameters for aerospace vehicles may change markedly when compressibility effects are experienced as flight speeds increase. Mach number, the ratio of aircraft velocity to the velocity of sound in the compressible fluid medium in which the aircraft is flying, is required to be the same for aircraft and model for similitude when compressibility effects dominate. Although the Mach number based on the speed of the complete aircraft may be low, in some cases, significant compressibility effects may be present on components of the vehicle. For example, flow over the upper surface of a wing with a thick high-lift airfoil at high angles of attack may experience Mach numbers approaching 1, even though the aircraft's flight speed may be as low as Mach 0.3. As the local flow approaches transonic Mach numbers, severe flow separation can occur because of interactions of shock waves formed on the wing and the wing's boundary layer flow. In turn, flow separation can cause undesirable dynamic motions such as wing drop or instability in pitch. When Mach number effects are expected to dominate, another set of scaling laws must be used for compressible-flow conditions:

SCALE FACTOR	
Linear dimension	n
Relative density ($m/\rho l^3$)	1
Mach number	1
Froude number (V^2/lg)	1
Angle of attack	Dependent on Froude scaling
Linear acceleration	1
Weight, mass	n^3/σ
Moment of inertia	n^5/σ
Linear velocity	$n^{1/2}$
Angular velocity	$1/n^{1/2}$
Time	$n^{1/2}$
Reynolds number (Vl/v)	$n^{1.5}v/v_0$

Scale factors for rigid dynamic models when matching Mach number. Multiply full-scale values by scale factors to determine model values, where n is the ratio of model-to-full-scale dimensions, σ is the ratio of air density to that at sea level (ρ/ρ_0), and v is the value of kinematic viscosity.

To maintain equivalent Mach numbers between model and aircraft, the previously discussed similitude requirements for Froude number (V^2/lg) can no longer be met without an unfeasible change in the gravitational field. One result of this deficiency is that the model must be flown at a different angle of attack than the full-scale vehicle. The model will be at a lower angle of attack than the full-scale aircraft while flying at the same Mach number. With this discrepancy in mind, NASA's applications and experiences with free-flying models for dynamic motion studies have included incompressible and compressible conditions with properly scaled models.[4]

Limitations and Interpretation of Results

Unfortunately, all the similitude requirements previously discussed cannot be duplicated in every respect during model testing. Use of the dynamic model test techniques therefore requires an awareness of the limitations inherent to the techniques and an accumulation of experience in the art of extrapolating results to conditions that cannot be directly simulated. In addition, results should not be extended beyond their intended areas of application.

Unquestionably, the most significant issue involved in free-flight testing is the discrepancy in values of Reynolds number between the model and full-scale vehicle.[5] Most of the critical flight issues studied by NASA in dynamic free-flight model tests involve partially or fully separated flow conditions, which may be significantly affected by Reynolds number. Reynolds number effects are generally configuration dependent and may be very complex.

One of the more common Reynolds number effects of concern involves its potential effects on the variation of the magnitude and angle of attack for maximum lift. Shown in the accompanying figure are lift data measured in wind tunnel tests of a general-aviation model at values of Reynolds number representative of those used in dynamic free-flight model tests and Reynolds numbers representative of full-scale flight.

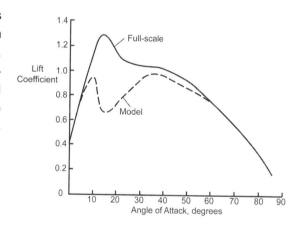

The data show significant increases in the magnitude and angle of attack for maximum lift as Reynolds number increases from model conditions to the full-scale value. Such results can significantly affect the prediction of flight characteristics of the airplane near and above stall. For example, the trends shown by the data might significantly influence critical characteristics, such as spin resistance. When an aircraft is flown near the angle of maximum lift, any disturbance that results in one wing dropping will subject that wing panel to a local increase in angle of attack. If the variation of lift with angle of attack is negative (as it normally is

Effect of Reynolds number on lift of a general-aviation model as measured in wind tunnel tests. Data were obtained at relatively low values of Reynolds number corresponding to model free-flight tests and high values representative of full-scale flight tests. The full-scale data show a markedly higher magnitude and angle of attack for maximum lift. Primarily a function of wing geometric features, such scale effects can lead to erroneous model results.

immediately beyond maximum lift), the down-going wing will experience a loss of lift, further increasing the tendency for the wing to drop and resulting in a pro-spin propelling phenomenon known as wing autorotation. If the configuration has large Reynolds number effects, such as those shown in the sketch, the extension of the angle of attack for maximum lift at full-scale conditions may delay the spin angle of attack for the airplane to a higher value. As a consequence, the model might exhibit a steeper spin from which recovery might be relatively easy, whereas the full-scale airplane might exhibit a more dangerous spin at higher angles of attack and unsatisfactory spin recovery.

Finally, a consideration of the primary areas of application of free-flight dynamic model testing is in order to provide a cautionary note regarding extrapolating free-flight results to other issues that are best analyzed using other engineering tools. In particular, caution is emphasized when attempting to predict pilot handling quality evaluations by using remotely piloted free-flight models. NASA's experience has shown that remotely controlled models do not provide suitable physical cues for the pilot when trying to quantitatively rate flying qualities. Instead, these studies are best conducted with the pilot situated in the simulator cockpit, with appropriate visual and motion cues provided at full-scale values of time. Investigations of handling characteristics for precision maneuvers are much more suitably conducted using fixed or moving-base piloted simulators.

With NASA's awareness and cautious respect for scale effects, its success in applying dynamic free-flight models for evaluations of conventional and radical aerospace configurations over the years has been extremely good. Thousands of configurations have been tested to date using several techniques, and the results of many of the NASA investigations have included correlation with subsequent flight tests of full-scale vehicles. Subsequent sections provide overviews of some of the success stories involved in this research as well as notable lessons learned.

CHAPTER 2:

HISTORICAL DEVELOPMENT BY THE NACA AND NASA

Early pioneers of aeronautical engineering were overwhelmed with the complexity of technologies required for the design and analysis of successful aircraft configurations. Long before the advent of high-speed digital computers and other sophisticated methods routinely used today, experimental methods in the areas of aerodynamics, structures and materials, propulsion, and the dynamics of flight were the primary tools of the design community. Subscale models provided basic understandings of fundamental phenomena as well as data for correlation with analytical studies or full-scale flight results. The following discussion provides highlights of the development of testing techniques directed at the study of flight dynamics, with an emphasis on dynamic stability and control. The development of ground-based facilities for dynamic model tests will be covered first, followed by a review of outdoor testing techniques.

At the Langley Memorial Aeronautical Laboratory of the National Advisory Committee for Aeronautics, applications of aircraft models for conventional wind tunnel testing in the 1920s were quickly followed by applications of specialized free-flight models for a broad variety of research investigations involving the prediction of stability and control characteristics, dynamic motions during controlled and uncontrolled flight, structural responses to atmospheric gusts and turbulence, and landing loads. As the merits of dynamic model testing became established and appreciated, specialized facilities and testing techniques were quickly advocated, developed, and put into operational status. With a dedicated and brilliant young staff of researchers, Langley's management provided the encouragement and resources required for the development of research facilities that would ultimately serve the interests of the Nation far beyond those that were initially envisioned. As a result of decades of continual commitment to the technical subject, Langley's experiences have been the focal point for NASA's contributions using dynamic free-flight models.[1] Critical contributions and accomplishments have also been made at the Dryden and Ames Research Centers, as will be discussed.

Spin Research Facilities

One of the first life-threatening problems of heavier-than-air flying machines demanded early attention from designers and pilots of the day—the dreaded tailspin. The international aviation community knew very little about the critical factors that influenced the spin behavior of aircraft or the relative effectiveness of different piloting methods to terminate the spin and recover to conventional flight. Langley researchers were the first international group to use dynamically scaled models for spin research.[2] They initially evaluated the use of catapult-launched models with wingspans of about 3 feet in an Army airship hangar about 105 feet high at Langley Field. After a series of free-flight studies of spin entries, the Langley staff concluded that the catapult technique was fundamentally limited and prone to operational problems. The approach was very time-consuming and resulted in frequent damage to the fragile models, and the abbreviated flight time did not permit the fully developed spin to be entered before the model impacted the recovery net. In addition, critics questioned the validity of the subscale model tests because of potential aerodynamic scale effects.

Meanwhile, British researchers at the Royal Aircraft Establishment (R.A.E.) had been inspired by the NACA efforts and began their own assessments of a catapult technique. Arriving at similar conclusions as their American peers regarding the limitations of the catapult technique, they developed a new free-drop technique in a large building.[3] To permit more time to study developed spins before impact, they constructed a special model-launching mechanism that used a threaded vertical rod to launch models with prespin motions from the top of a balloon shed. After the balloon shed was demolished during a planned expansion of facilities at Farnborough, U.K., the R.A.E. built a small vertical wind tunnel with a 2-foot test section to determine if a free-spinning model test technique would be feasible. The exploratory tests were a huge success and led to the construction of a 12-foot-diameter vertical spin tunnel at Farnborough, which was put into operation in 1932.[4]

At Langley, efforts to promote an understanding of the aerodynamics of the spin had resulted in the construction of a 5-Foot Vertical Tunnel in the late 1920s to provide for measurements of the aerodynamic loads on aircraft models during simulated spinning motions.[5] Airflow in this early tunnel was vertically downward to simplify the balancing of mass loadings during rotary motions, and the model was rigidly mounted to a rotating balance assembly driven by an electric motor. Aerodynamic data from the tests were then used in theoretical analyses of potential spin modes that might be exhibited by the full-scale airplane. A considerable amount of valuable data was gathered in the 5-Foot Vertical Tunnel, but the advantages of free-spinning test techniques were obvious. Aware of the success of the British free-spinning tunnel at Farnborough, Langley then proceeded to design and construct a similar 15-foot free-spinning tunnel in 1934 that permitted researchers to observe the spin and recovery motions of small free-flying balsa models.[6] Operations began in 1935 as the first operational vertical free-spinning tunnel in the United States. The designer, Charles H. Zimmerman, had been a key member of the staff of the 5-Foot Vertical Tunnel. Zimmerman was a brilliant researcher and innovative tinkerer, who developed several dynamic model testing techniques and radical new airplane concepts.

The airflow in the 15-Foot Spin Tunnel was vertically upward to simulate the downward velocity of an aircraft during spins. At the beginning of a typical test, the model was mounted on a launching spindle at the end of a long wooden rod held by a tunnel technician. A tunnel operator increased the airspeed in the vertical tunnel so that the air forces on the model equaled its weight. At this point, the model automatically disengaged from the spindle and continued to float as the airspeed was continually adjusted to maintain the model's position at eye level in the test section. The model's control surfaces were moved from pro-spin settings to predetermined anti-spin settings by a clockwork mechanism, and the rapidity of recovery from the spin was noted. After the test was completed, the tunnel airspeed was decreased, and the model was allowed to descend and settle into a large net at the bottom of the test section. The model was then recovered with a long-handled clamp and prepared for the next test. After several years of operation, the launching spindle was discontinued, and the model was merely launched by hand into the airstream with an initial spinning motion.

Free-spinning test underway in the Langley 15-Foot Spin Tunnel. The researcher on the left has placed the model and its launching spindle in the vertically rising airstream with pro-spin settings on the control surfaces. Motion-picture records were made after the model reached a steady spinning condition. Spin recovery controls were actuated by a clockwork mechanism within the model. Measurements were made of the rate of spin, attitude of the model, and the number of turns required for recovery from the spin.

The Langley 20-Foot Spin Tunnel

In 1941, Langley replaced the 15-Foot Spin Tunnel with a new 20-foot tunnel, which has been in continual use to the present day. The facility features are similar to its predecessor, but it is capable of much higher

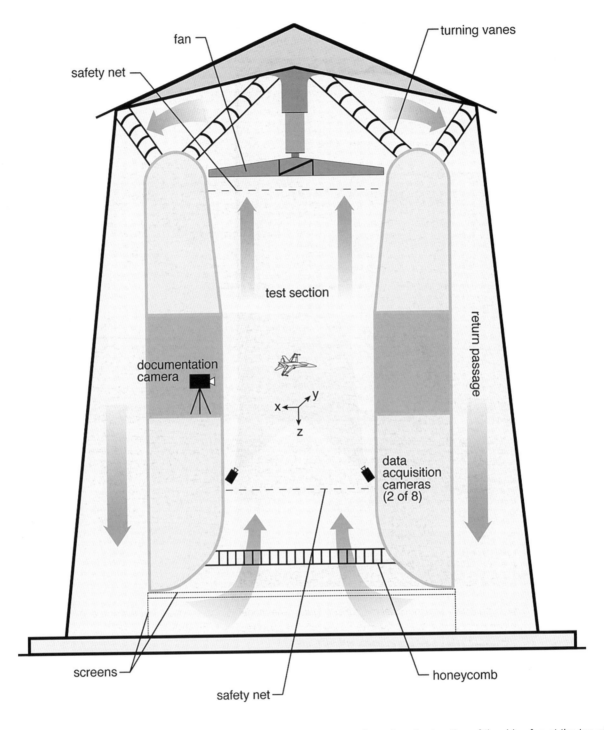

fan

turning vanes

safety net

test section

return passage

documentation
camera

y

x

z

data
acquisition
cameras
(2 of 8)

screens

safety net

honeycomb

Cross-sectional view of the Spin Tunnel shows the closed-return tunnel configuration, the location of the drive fan at the top of the facility, and the locations of safety nets above and below the test section to restrain and retrieve models. The tunnel operator uses a rapid-response airspeed control to maintain the position of the model within the field of view of data acquisition cameras. The Spin Tunnel became operational in 1941 and has conducted spin and recovery studies for nearly 600 aircraft designs and low-speed dynamic stability tests of numerous spacecraft.

Spin tunnel test of a dynamically scaled model of the F-22 fighter. The model has been hand-launched with a spinning motion into a vertical airstream controlled by the tunnel operator on the left. The spin recovery controls will be applied using the hobbyist-type, radio-controlled transmitter held by the technician on the right.

speeds for tests of larger, heavier models of current-day aircraft. The free-flying unpowered aircraft models are hand-launched to evaluate spinning and spin recovery behavior, tumbling resistance, and recovery from other out-of-control situations. The tunnel is a closed-throat, annular return wind tunnel with a 12-sided test section 20 feet across by 25 feet in length. The test section airspeed can be varied from 0 to approximately 85 feet per second. Airflow in the test section is controlled by a 3-bladed fan powered by a 400-horsepower direct current motor in the top of the facility, and the airspeed control system is designed to permit rapid changes in fan speed to enable precise location of the model in the test section.[7]

The models for free-spinning tests in the Langley 20-Foot Spin Tunnel are hand launched with a spinning motion into the vertically rising airstream and a tunnel operator controls the airspeed to stabilize the spinning model in view of observers. Data measured include attitude, spin rate, and control positions. For many years, data acquisition was accomplished by laborious manual indexing of motion-picture records taken during the test sequence. Over time, the acquisition system has become much more efficient and sophisticated. Today's system is based on a model space positioning system that uses retro-reflective targets attached on the model for determining model position.[8] Spin recovery controls in early operations were actuated at the operator's discretion rather than the preset clockwork mechanisms of the previous tunnel. Copper coils placed around the periphery of the tunnel set up a magnetic field in the tunnel when energized, and this magnetic field actuated a magnetic device in the model to operate the control surfaces. Today, the control surfaces are actuated using premium off-the-shelf radio-controlled hobby gear.

This versatile testing facility is also used to determine the size and effectiveness of emergency spin recovery parachute systems for flight-test aircraft by deploying various parachute configurations from the spinning model. In addition, unconventional maneuvers such as uncontrollable nose-over-tail gyrations known as "tumbling" can be investigated. In support of NASA's space exploration programs, the tunnel has also been used to evaluate the stability of vertically descending configurations such as parachute/capsule systems. In addition to its capabilities for free-flight testing, the Spin Tunnel is equipped with special test equipment to permit measurement of aerodynamic forces, moments, and surface pressures during simulated spinning motions, as will be discussed in a later section. In recent years, the capability of obtaining such measurements during combined rotary and oscillatory motions has also been added to the research equipment used in the facility. The Spin Tunnel has supported the development of nearly all U.S. military fighter and attack aircraft, trainers, and bombers during its 68-year history, with over 600 projects to date.

After the introduction of vertical spin tunnels for aeronautical research in England and the United States, widespread developments of similar facilities occurred worldwide, including free-spinning tunnels in France, Canada, Australia, Germany, Italy, Russia, China, and Japan.

Wind Tunnel Free-Flight Research Facilities

After the successful introduction of Zimmerman's 15-Foot Spin Tunnel in 1935, Langley researchers extended their visions to include a wind tunnel method to study the dynamic stability and control characteristics of a free-flying aircraft model in conventional flight. This concept consisted of placing an unpowered, remotely controlled model within the test section of a "tilted" wind tunnel, increasing the airspeed until the aerodynamic lift of the model supported the weight of the model, and then observing the model's stability and response to control inputs as it flew in gliding flight. Zimmerman first conceived and fabricated a 5-foot-diameter proof-of-concept free-flight wind tunnel that was suspended from a yoke, which permitted it to be rotated about a horizontal axis. The closed-throat, open-return atmospheric tunnel was capable of being tilted from the horizontal position to a maximum nose-down orientation of 25 degrees. No provision was made for climb angles, because no test of powered models in the tunnel was contemplated.

The tilt angle of the 5-Foot Free-Flight Tunnel and the airspeed produced by a fan at the rear of the test section were controlled by a tunnel operator at the side of the test section. An evaluation pilot at the rear of

Free-flight tests of a model of the Brewster Buffalo airplane in the 5-Foot Free-Flight Tunnel. The researcher on the left controls the tilt angle and airspeed of the yoke-mounted tunnel, while the evaluation pilot on the right provides control inputs via light flexible cables from his control box to the model.

the tunnel provided inputs to the model's controls via fine wires that were kept slack during the flight. Meanwhile, the tunnel operator adjusted the airspeed and tunnel angles so that the model remained relatively stationary near the center of the tunnel during a test. By coordinating their tasks, the two researchers were able to assess the dynamic stability and control characteristics of the test model. The information obtained in the test was qualitative and consisted primarily of ratings for various stability control characteristics based on observations of a pilot and tunnel operator.

The models tested in the 5-foot tunnel were quite small (wingspan approximately 2 feet) with very light construction (balsa shell or balsa framework covered with paper). Small electromagnetic actuators were used to deflect the control surfaces, and power to operate the actuators was supplied through light flexible wires that trailed freely from the model to the floor of the tunnel. As a result of the relatively small size of the dynamically scaled model, the motions of the models were very fast and difficult to control.

In early testing, the research team lacked the coordination and experience required for satisfactory studies, and inadvertent crashes of the fragile models curtailed many research programs when the facility began operations in 1937. In support of the development of the free-flight capability, Langley's instrumentation experts developed an automatic light-beam-control device that stabilized the models and greatly improved the training of research pilots and tunnel operators.

The Langley 12-Foot Free-Flight Tunnel

Only a few research projects were conducted in the 5-Foot Free-Flight Tunnel, but they were quite successful and stimulated interest in a more capable research facility. Charles Zimmerman already had his design in mind for an even larger, more capable tunnel. A 12-foot free-flight tunnel was soon constructed and placed into operation in 1939. The new tunnel had an octagonal test section that was 15 feet in length, and the length of the tilting portion of the tunnel was 32 feet. The Langley 12-Foot Free-Flight Tunnel was housed in a 60-foot-diameter sphere so that the return passage for the air would be essentially the same for all tunnel tilt angles. Powered models were under consideration, so the tilt angle of the test section ranged from 40-degree glide to 15-degree climb conditions. Three operators were used during tests. Two of the researchers were positioned at the side of the test section to control tunnel angle and airspeed. One of the operators also controlled propulsion power to the model for power-on flight tests. The pilot was enclosed in a viewing area at the rear of the test section and controlled the model with inputs to two small control sticks, which activated small electromagnetic control servos in the model.[9]

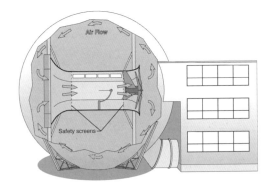

Sketch of the 12-Foot Free-Flight Tunnel and its adjacent office building. The 60-foot-diameter sphere housing the facility provided a continuous return passage for the closed-return tunnel airstream as the tunnel pivoted for free-flight tests.

Free-flight test of a swept-forward wing, tailless fighter design known as the Air Material Command MCD-387-A in the Langley 12-Foot Free-Flight Tunnel. The evaluation pilot is visible in the enclosure at the lower left, while the two tunnel operators controlling airspeed and tunnel tilt angle are at the right.

Models designed for tests in the 12-foot tunnel had wingspans of about 3 feet and were constructed of solid balsa wings and hollow balsa fuselages in early tests, but later tests used models of built-up construction of spruce and other materials. Tests of simplified research models used aluminum-alloy fuselage booms on many occasions. The control actuators used by the flight models were spring-centered electromagnetic mechanisms that were later succeeded by solenoid-operated air valves. One of the major lessons learned in early free-flight experiences with the smaller dynamically scaled models was that full-on/full-off "bang-bang" controls were much more satisfactory than conventional proportional-type controllers. It was found that the rapid angular motions of the small models required large control effectiveness to maintain adequate control while studies were made of the inherent motions. Artificial rate dampers, consisting of small air-driven rate gyroscopes interconnected with proportional-type pneumatic control actuators, were developed and used for artificial damping.

After operations began in the tunnel, an external-type aerodynamic balance system was implemented above the ceiling of the test section for measurements of aerodynamic performance, stability, and control characteristics of the free-flight models. In later years, state-of-the-art internal electronic strain-gauge balances were used in test programs.

Located near the Spin Tunnel, the Langley 12-Foot Free-Flight Tunnel (now known as the Langley 12-Foot Low-Speed Tunnel) provided a unique free-flight test facility for national programs for over 15 years. However, researchers involved in free-flight model tests were continually looking for a larger facility to permit more space for flight maneuvers of larger models that were more representative of full-scale aircraft. The free-flight tunnel staff had even successfully conducted free-flight tests of VTOL models as early as 1956 within the gigantic 30- by 60-foot test section of the Langley Full-Scale Tunnel. Those tests were soon followed by a series of flight tests of other airplane configurations. In 1959, two events occurred that shifted the site of free-flight testing at Langley. First, a slack in the post–World War II operational schedule of the Full-Scale Tunnel permitted more free-flight tests to be conducted with larger, more sophisticated models, more spatial freedom, and less risk during flight evaluations. The second event was the organization and staffing of the new NASA Space Task Group (STG) at Langley to prepare for Project Mercury. Many staff members from the Full-Scale Tunnel transferred to the STG and moved to other locations, resulting in a significant depletion of the tunnel staff. Consequently, the free-flight tunnel staff and their model flying techniques were reassigned to the larger wind tunnel. The 12-Foot Free-Flight Tunnel was subsequently converted for conventional aerodynamic force-test studies of powered and unpowered models, and its test section was fixed in a horizontal attitude. The tunnel has been in continual use by NASA Langley for over 50 years and has proven to be a valuable low-cost, rapid-access asset for early assessments of radical new concepts and configurations before more extensive tests are scheduled in other tunnels.

The pioneers who led the early development and applications of dynamic model flight efforts in the various spin tunnels and free-flight tunnels prior to 1960 became scientific legends within the NACA and NASA. Individuals such as Charles H. Zimmerman, Charles J. Donlan, Joseph A. Shortal, Hubert M. Drake, Hartley A. Soule, Gerald G. Kayten, and John P. Campbell served long and distinguished careers with the NACA into the NASA years, during which they became famous figures in NASA's most historic aeronautical and space programs.

The Langley Full-Scale Tunnel

The move to the Full-Scale Tunnel was a gigantic step in the development of the wind tunnel free-flight technique and the construction methods used for the free-flight models. This large tunnel became the centerpiece of free-flight studies at Langley because of its unique design features. The tunnel is closed-circuit and open-throat, characterized by an open quasi-elliptical test section with dimensions of 60 feet across by 30 feet high and a length of 56 feet. The air is drawn through the test section at speeds up to about 80 mph by two 4-bladed, 35.5-foot-diameter propellers powered by two 4,000-horsepower electric motors. After the airstream passes through the test section, it returns to it in a bifurcated manner through wall passageways within the building enclosing the test section.

The huge Langley Full-Scale Tunnel has been the focal point of NASA's free-flight model studies. The large test section and open-throat design permit flight-testing of relatively large sophisticated models of aerospace vehicles.

For free-flight investigations, pilots fly the remotely controlled, powered model with minimal restraint in the 30- by 60-foot open-throat test section of the tunnel for various test conditions and airspeeds. Flying the model involves a carefully coordinated effort by the test team, whose members are at two sites within the wind tunnel building. One group of researchers is in a balcony at one side of the open-throat test section, while a pilot who controls the rolling and yawing motions of the model is in an enclosure at the rear of the test section within the structure of the tunnel exit-flow collector. Compressed air powers the model, and a thrust pilot in the balcony controls the level of thrust. Seated next to the thrust pilot is the pitch

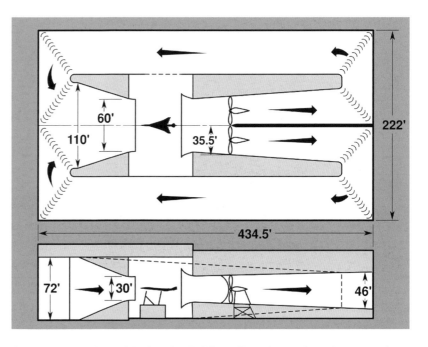

Cross-sectional views of the Langley Full-Scale Tunnel show the unique open-throat test section and large dimensions of the facility. The arrangement permitted pilots and crews to be outside the airstream during free-flight operations. With a continual airspeed control from about 20 mph to about 80 mph and the ability to withdraw the model from the test section in the event of loss of control, the tunnel provided an ideal test site for flying models.

pilot, who controls the longitudinal motions of the model and conducts assessments of dynamic longitudinal stability and control during flight tests.

The requirement for multiple pilots because of the impact of dynamic scaling of models has already been discussed. In addition to the model's high angular rates, the pilot is limited by flying the model remotely with only visual cues, and he cannot sense the cues provided by the accelerations of the model—in contrast to a pilot onboard the full-scale airplane. The lack of acceleration cues can result in delayed pilot inputs and pilot-induced oscillations, which can be critical in the relatively constrained area of the tunnel test section. Other key members of the test crew include the test conductor and the tunnel airspeed operator.

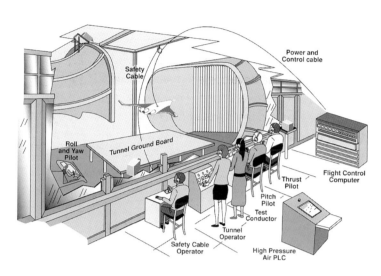

Test setup for free-flight studies in the Langley Full-Scale Tunnel. The pitch pilot and other operators concerned with the longitudinal motions of the model and its horizontal and vertical positions in the tunnel test section are in a balcony at the side of the test section, out of the airstream. The pilot who controls the rolling and yawing motions of the model is seated in an enclosure at the rear of the tunnel, behind the model. A digital flight control computer is used to simulate control laws.

Free-flight evaluation of the dynamic stability and control of a powered model of the YF-22 fighter prototype at high angles of attack in the Langley Full-Scale Tunnel. The model was powered with compressed air ejectors and included thrust-vectoring engine nozzles and a nose boom equipped with angle-of-attack and sideslip vanes and sensors for implementation of critical flight control system elements of the full-scale aircraft. The outstanding behavior of the model at high angles of attack was subsequently verified by the airplane.

A light, flexible cable attached to the model serves two purposes. The first is to supply the model with compressed air, electric power for control actuators, and transmission of feedback signals for the controls and sensors carried within the model. The second purpose involves safety. A portion of the cable is made up of steel that passes through a pulley above the test section. This part of the flight cable is used to snub the model when the test is terminated or when an uncontrollable motion occurs. The entire flight cable is kept slack during the flight tests by a safety-cable operator in the balcony, who accomplishes this job with a high-speed winch.

The introduction of larger models and the emergence of advanced flight control systems for full-scale aircraft posed significant challenges for the design, fabrication, and instrumentation of free-flight models and the Full-Scale Tunnel. Model dimensions greatly increased as model wingspans approached 6 feet or greater, scaled weights for the models approached 100 pounds, and new types of propulsion units such as miniature turbofans and high thrust/weight propeller propulsion units required development. Fabrication of models changed from the simple balsa free-flight construction used in the 12-Foot Free-Flight Tunnel to high-strength, lightweight composite materials. The development and implementation of new model structures compatible with the rigors of flight-testing and loads imposed during severe instabilities were made even more complicated by the requirements of maintaining accurate scaling relationships for weight and mass distribution of the models.

In recent years, the control systems used by the free-flight models have been upgraded to permit simulation of the complex feedback and stabilization logic involved in flight control systems for contemporary aircraft. The control signals from the pilot stations are transmitted to a digital computer in the balcony, and a special software program computes the control surface deflections required in response to pilot inputs, sensor feedbacks, and other control system inputs. Typical sensor packages include control-position indicators, linear accelerometers, and angular-rate gyros. Many models employ nose-boom-mounted vanes for feedback of angle of attack and angle of sideslip, similar to systems used on full-scale aircraft. Data obtained from the flights include optical and digital recordings of model motions, as well as pilot comments and analysis of the model's response characteristics.

The free-flight technique is used to obtain critical information on the behavior of aerospace configurations in 1 g level flight from low-speed stall conditions at high angles of attack to high-speed subsonic conditions at low angles of attack. Specific objectives are to evaluate the configuration's dynamic stability and response to control inputs for flight at several angles of attack, up to and including conditions at which loss of control occurs at extreme angles of attack. In addition, an assessment is made of the effectiveness of the flight control system feedbacks and augmentation systems as well as airframe geometric modifications.

A typical test sequence begins with the model ballasted for a specific loading. The model initially hangs unpowered from the safety cable and is lowered into the test section. Tunnel airspeed is then increased to the condition of interest, at which the model is trimmed with thrust and control inputs from the pilots and flies without restraint for the specified flight conditions.

The technique has proven to be excellent for obtaining qualitative evaluations of the flying characteristics of aircraft configurations, especially unconventional and radical designs for which experience and data are lacking or nonexistent. Because the tests are conducted indoors, the test schedule is not subject to weather, and a relatively large number of tests can be conducted in a short time. Configuration modifications are quickly evaluated, and the models used in the flight tests are also used in conventional wind tunnel force and moment tests, which provide aerodynamic data for use in the analysis of the observed flight motions and the development of mathematical models of the configuration for use in piloted simulators.

For over 50 years, the Langley Full-Scale Tunnel has been the site for a continual stream of free-flight investigations of aircraft and spacecraft configurations. As will be discussed in later sections, configurations studied have included high-performance fighters, vertical/short take-off and landing (V/STOL) aircraft, parawing vehicles, advanced supersonic transport configurations, general aviation designs, new military aircraft concepts, and lifting bodies. Now under the operation of Old Dominion University (ODU) of Norfolk, VA, the Full-Scale Tunnel continues to support aeronautics research and development to this day.

The Ames Hypervelocity Free-Flight Facilities

In addition to its stable of low-speed dynamic model free-flight facilities, NASA has developed specialized ground-based free-flight testing techniques to determine the high-speed aerodynamic characteristics, static stability, and dynamic stability of aircraft, Earth atmosphere entry configurations, planetary probes, and aerobraking concepts. The NASA Ames Research Center led the development of such facilities starting in the 1940s with the Ames Supersonic Free-Flight Tunnel (SFFT).[10] The SFFT, which was similar in many respects to ballistic range facilities used for testing munitions, was a blow-down facility specifically designed for aerodynamic and dynamic stability research at high supersonic Mach numbers. Test speeds extended from low supersonic speeds to Mach numbers in excess of 10. In this unique facility, the model under test was fired at high speeds upstream into a supersonic wind tunnel airstream (typically Mach 2). The 1-foot by 2-foot test section of the tunnel was 18 feet long and used the Ames 12-Foot Low-Turbulence Pressure Tunnel as a reservoir. Windows for shadowgraph photography were along the top and sides of the test section. The free-flight models were launched from guns about 35 feet downstream of the test section. The launching process included the use of special sabots that transmitted the propelling forces of the gun to the model during launch, after which the sabots aerodynamically separated from the model. The model's flightpath down the test range achieved high test Reynolds numbers.

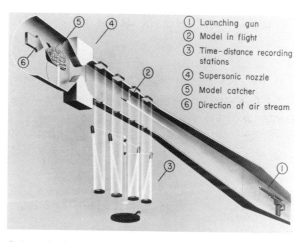

① Launching gun
② Model in flight
③ Time-distance recording stations
④ Supersonic nozzle
⑤ Model catcher
⑥ Direction of air stream

Schematic drawing of the Ames Supersonic Free-Flight Tunnel.

Model, sabot, and cartridge assembly prepared for firing in the Ames Supersonic Free-Flight Tunnel.

Aerodynamic data were derived from motion time histories and measurements of the model's attitudes during the brief flights. Dynamic stability characteristics could also be observed for the test article. Optical approaches for data reduction are required, because surface-mounted sensors and telemetry electronics would be destroyed, along with the model, at the terminal wall of the range. The development of the test technique and the associated instrumentation required years of work by the dedicated Ames staff. For example, the small research models had to be extremely strong to withstand high accelerations during the launch (up to 100,000 g's) yet light enough to decelerate for drag determination while meeting requirements for moments of inertia. Launching the models without angular disturbances or damage required extensive development and experience. Another major issue facing the researchers was data contamination caused by oblique shock waves in the test section. To resolve this problem, the upper and lower walls of the test section were diverged to allow the flow to expand steadily and avoid discontinuous compression at the shocks.

View of the Physical Research Laboratory at Langley in 1949 showing the ill-fated Langley Free-Flight Facility in the foreground. The 100-foot-long, 8-foot-diameter facility was equipped with viewing windows but suffered poor flow characteristics during model flights and was closed before meaningful research results were obtained.

The Supersonic Free-Flight Tunnel was completed in late 1949 and became operational in the early 1950s.[11] The unique testing capabilities of this Ames facility provided valuable information on supersonic drag, lift-curve slope, pitching moment variations with angle of attack, aerodynamic damping in roll, and the location of center of pressure of test specimens.[12]

Efforts to develop a high-speed, free-flight apparatus had also taken place at Langley in the late 1940s at the Langley Physical Research Laboratory. The test apparatus consisted of an 8-foot-diameter tank with a length of 100 feet in which models would be propelled from a compressed-gas gun at speeds between 500 and 1,000 mph. The test mediums for the facility included air, Freon, and mixtures of the two gases. Unfortunately, early assessments of the free-flight test data obtained in the facility indicated severe choking and unacceptable aerodynamic contamination of results. In 1949, Langley moved its high-priority 11-Inch Hypersonic Tunnel to the same control-room site occupied by the free-flight facility and terminated the free-flight activities. The facility was then demolished, without producing significant technical output.

In 1958, free-flight aeroballistic testing evolved considerably with the advent of the Ames Pressurized Ballistic Range (PBR). This facility had a 203-foot-long, 10-foot-diameter test section with 24 orthogonal shadowgraph stations and could be operated at pressures ranging from 0.1 to 10 atmospheres (providing an impressive range of achievable Reynolds number variations). Models were launched from an arsenal of guns (powder and light-gas of various sizes) into this test section. Aerodynamic testing was typically performed at velocities up to 10,000 feet/second, whereas some ablation studies were conducted up to 22,000 feet/second. Some of the major programs and missions supported in the PBR were X-15, Mercury, Gemini, Polaris, Apollo, Viking, and NASA's Aero-assist Flight Experiment (AFE). The PBR was last operated in 1987.

As light-gas gun technology continued to evolve and velocity capabilities continued to increase, NASA brought online its most advanced aeroballistic testing capability, the Ames Hypervelocity Free-Flight Aerodynamic Facility (HFFAF), in 1964. This facility was initially developed in support of the Apollo program and utilized both light-gas gun and shock tube technology to produce lunar return and atmospheric entry conditions (24,600 mph relative model velocities). At the center of the facility was a 75-foot-long test section that could be evacuated to subatmospheric pressure levels or backfilled with various test gases to simulate flight in various planetary atmospheres. The test section was configured with 16 orthogonal shadowgraph imaging stations and a high-precision, spatial reference wire system. At one end of the test section, a family of light-gas guns (ranging in size from 0.28 to 1.50 caliber) was used to launch aerodynamic models (at speeds up to 13,400 mph) into the test section, while at the opposite end a large shock tube could be simultaneously used to produce a counterflowing airstream (the result being relative model velocities approaching 24,600 mph or Mach numbers of about 30). This counterflow mode of operation proved to be very challenging and was used for only a brief time from 1968 to 1971. Throughout much of the 1970s and 1980s, this versatile facility was operated as a traditional aeroballistic range, using the guns to launch models into quiescent air (or some other test gas), or as a hypervelocity impact test facility. From 1989 through 1995, the facility was operated as a shock tube–driven wind tunnel for scramjet propulsion testing. In 1997, the

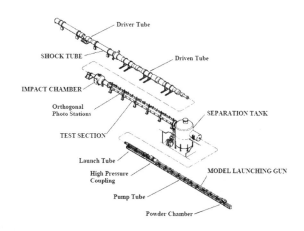

Sketch of the Ames Hypervelocity Free-Flight Aerodynamic Facility. Equipped with a shock tube for extreme Mach numbers, the facility has provided extensive supersonic and hypersonic free-flight data in a variety of investigations.

View of the Hypervelocity Free-Flight Aerodynamic Facility during recent tests in support of the NASA Orion project.

HFFAF underwent a major refurbishment and was returned to an aeroballistic mode of operation with extended light-gas gun capabilities and powder-gas guns that had been previously used in the PBR.[13] It continues to operate in this mode and is NASA's only remaining aeroballistic test facility.

In its current (circa 2009) operating configuration, the HFFAF utilizes a suite of guns (powder and light-gas) to propel models into the aerodynamic test section, wherein the shadowgraph imaging stations are used to capture the model's flight time history. The resultant trajectory record is then used to extract critical aerodynamic parameters for the configuration being studied. In addition, infrared cameras can be positioned at various stations to record model surface temperature distributions at different points along the model's flight path. From this information, heat transfer rates and transition to turbulence locations can be inferred. Some of the major programs and missions that have been supported in the HFFAF include Apollo, Viking, Pioneer Venus, Galileo, Shuttle, International Space Station, National Aero-Space Plane (NASP), Mars Science Laboratory, and the Crew Explorations Vehicle (CEV/Orion).[14] In addition, the HFFAF has been used frequently for fundamental aerodynamics testing, material testing, and sonic boom research.[15]

A one-quarter-scale free-flight model of the Lockheed XFV-1 VTOL tail sitter airplane being prepared for hovering tests in the return passage of the Ames 40- by 80-Foot Tunnel. The test of the 250-pound model was the only subscale free-flight test conducted in the facility. Note the propeller-guard assembly for the counter-rotating propellers, the flight cable providing power and control inputs, and the wingtip-mounted stabilization lines.

The Ames 40- by 80-Foot Tunnel

Ames researchers also briefly conducted an exploratory free-flight model test in the gigantic 40- by 80-Foot Tunnel in the early 1950s. Lockheed and Ames partnered for an exploratory investigation to determine the hovering and low-speed flight characteristics of a large powered model of the Lockheed XFV-1 "tail sitter" VTOL airplane.[16] The objectives of the free-flight activity were to demonstrate that the configuration was capable of successfully performing its operational requirements, to verify and supplement predictions of stability and control, and to assess the feasibility of piloting techniques in the 40- by 80-Foot Tunnel. Flying the 250-pound model in the closed test section of the tunnel remotely proved to be challenging for a single pilot. Details of the test setup and results will be discussed in a later section.

In recognition of the 40- by 80-foot facility's status as the premier NACA–NASA large-scale subsonic wind tunnel, the demands for conventional large-scale performance tests of aircraft and rotorcraft there precluded the possibility of conducting other subscale free-flight tests. To the author's knowledge, this test program has been the only free-flight model study conducted in the facility. Virtually all large-scale subsonic free-flight testing has been more suitably been directed to the Langley Full-Scale Tunnel.

Outdoor Free-Flight Facilities and Ranges

Although the foregoing wind tunnel free-flight testing facilities provide unique and valuable information regarding the flying characteristics of advanced aerospace vehicles, they are inherently limited or unsuitable for certain applications. In particular, vehicle motions for other than 1 g flight involving large maneuvers or out-of-control conditions result in significant changes in flight trajectory and altitude, which can only be studied in the expanded spaces provided by outdoor facilities. In addition, research directed at high-speed dynamic stability and control problems could not be conducted in Langley's low-speed wind tunnels. Outdoor testing of dynamically scaled free-flight models was therefore developed and applied in many research activities at Langley, Dryden, Ames, and Wallops.

Langley Unpowered Drop Models

NASA has used unpowered drop models for a variety of research and development activities involving advanced aircraft and space vehicles. Although these outdoor test techniques are more expensive than are wind tunnel free-flight tests and are subject to limitations because of weather, the results obtained are unique, cannot be obtained in wind tunnels, and are especially valuable for certain types of flight dynamics studies.

One of the most important drop-model applications is in the study of aircraft spin entry motions, which includes analyses of spin resistance and poststall gyrations. A significant void of information exists between the prestall and stall-departure results produced by the wind tunnel free-flight test technique in the Full-Scale Tunnel discussed earlier and the results of fully developed spin evaluations obtained during spin tunnel tests. This information can be critically misleading for some aircraft designs. For example, some aircraft configurations exhibit severe instabilities in pitch, yaw, or roll at stall during wind tunnel free-flight tests, and they may also exhibit potentially dangerous spins, from which recovery is impossible during spin tunnel tests. However, a combination of aerodynamic and inertial properties can result in this same configuration exhibiting a high degree of resistance to enter the dangerous spin following a departure, despite forced spin entry attempts by a pilot. On the other hand, some configurations easily enter developed spins without prolonged assistance from the pilot. Of equal importance, drop-model testing is required to define the most effective control strategies to be used by the pilot of the full-scale aircraft for recovery from out-of-control motions.

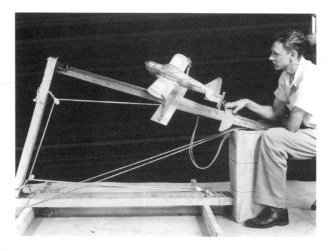

To evaluate the resistance of aircraft to spins, in 1950, NACA Langley revisited the catapult techniques of the 1930s and experimented with an indoor catapult launching technique within the same building (about 70 feet square and 60 feet high) that had earlier housed the 15-Foot Spin Tunnel and the 5-Foot Free-Flight Tunnel.[17] An

Free-flight model of a representative general-aviation airplane being prepared for spin entry tests using a catapult device in 1950. The launching apparatus used an elastic bungee cord to propel the model with preset controls into a stall and incipient spin.

elastic bungee cord launched the unpowered model from a launching platform near the ceiling of the building with preset fixed spin entry control settings. Data from the brief flight were recorded with motion pictures, and no instrumentation was carried within the model. A large net stretched across the bottom of the building was used to recover the model after the test flight. This exploratory study was stimulated by the growing requirement to define the spin resistance of radical new configurations of the 1950s and to develop a capability to extend the wind tunnel free-flight tests. Once again, however, the catapult technique proved to be much too cumbersome and limited by testing space, and other approaches to study spin entry were pursued. Although the catapult-launched model was not a primary test technique during modern times, NASA later used it in response to a quick-reaction request from the U.S. Air Force for analysis of spin entry characteristics of the B-58 aircraft, as will be discussed in the next chapter. These tests were conducted in a large airship hangar at the Weeksville Naval Air Facility at Elizabeth City, NC.

Disappointed by the inherent limitations of the catapult-launched technique, the Langley researchers began to explore the feasibility of an outdoor drop-model technique in which models could be launched from a helicopter at much higher altitudes, permitting more time to study the spin entry and the effects of recovery controls. Initially, limited experiments were conducted at Langley Air Force Base in a constrained area to evaluate various schemes for ground-based remote control of the models and methods for coordination of the operation. Encouraged by initial successes, researchers conducted a search to locate a more feasible test site for routine operations. The primary parameters for suitable test sites were a location near Langley and a large drop range away from human inhabitants and dwellings. After additional trials involving model launches from a helicopter at altitudes as high as 2,000 feet at the Patrick Henry Airport in Newport News, VA, the research operations began in 1958 at a low-traffic airport near West Point, VA, about 40 miles from Langley.

As testing progressed at West Point, the technique evolved into an operation consisting of launching the unpowered model at an altitude of about 2,000 feet and evaluating its spin resistance with remotely located,

ground-based pilots who attempted to promote spins by various combinations of control inputs and maneuvers. At the end of the evaluations, an onboard recovery parachute was deployed and used to recover the model. The model was retrieved after a ground landing. This test approach proved to be the prototype of an extremely successful testing technique that NASA updated and applied for over 50 years.[18]

Ground control stations for pitch and roll-yaw pilots in early drop model tests. The crew consisted of a pitch pilot, a roll-yaw pilot, and a tracker operator. Binoculars were used for close-in views of the model during flight, and a tracking camera recorded motion pictures of the model's motions.

The drop models used in studies of the mid-1950s represented a tremendous leap in sophistication for free-flight models of the time, as well as significant challenges for model designers and fabricators. The models were large (typically having fuselages of about 7 feet and weighing over 100 pounds) and were constructed of composite materials to withstand the high g's of ground impact, even when slowed by the recovery parachute. Forward-looking miniature motion-picture cameras were mounted on board the models in flowthrough engine inlets to photograph the positions of flow-direction vanes mounted on a nose boom as well as control position indicator lights mounted on the fuselage interior. The helicopter used for launches at the time experienced considerable turbulence under its fuselage at low forward speeds; therefore, the drop model was mounted at the end of a 5-foot-long extensible vertical shaft and lowered to a position below the helicopter fuselage for launch at a forward airspeed of about 60 knots.

Initially, two separate tracking units consisting of modified power-driven antiaircraft gun trailer mounts were used by two pilots and two tracking operators to track and control the model. One pilot and tracker were to the side of the model's flight path, where they could control the longitudinal motions following launch, while the other pilot and tracker were about 1,000 feet away, behind the model, to control lateral-directional motions. As the technique was refined in later years, a single dual gun mount arrangement was used by both pilots and a single tracker operator.

Drop model of the F-4 fighter mounted on launch helicopter during an evaluation of spin resistance in 1960 at the West Point, VA, airport. After the altitude and speed for launch conditions were satisfied, the model was lowered on a vertical shaft and released.

The success of drop-model testing in predicting spin resistance and spin recovery procedures for military aircraft was disseminated within appropriate Department of Defense (DOD) organizations, resulting in an immediate increase in workload for emerging configurations and an urgency to conduct additional efforts to locate a test site closer to the Langley Research Center for improved efficiency and turnaround time in research projects. Another undesirable characteristic of the West Point operation was the presence of concrete runways, which caused significant damage to models during some landings. In 1959, the Air Force granted Langley approval to conduct drop tests at the abandoned Plum Tree bombing range near Poquoson, VA, about 5 miles from Langley. The Air Force had cleared the marshy area under consideration of depleted bombs and munitions. Access to the property was negotiated with private citizens, and clearance for the NASA drop operations was monitored and approved by the control tower at Langley Air Force Base (the area was within the flyover zone of aircraft operations at Langley). A temporary building and concrete landing pad for the launch helicopter were added for operations at Plum Tree, and a surge of request jobs for U.S. high-performance military aircraft in the mid- to late 1960s (F-14, F-15, B-1, F/A-18, etc.) brought a flurry of test activities that continued until the early 1990s.

Aerial view of the NASA test site at Plum Tree and its proximity to Langley Air Force Base. The town of Poquoson, VA, is at the right of the photograph. The marshy, soft land was well-suited for drop-model testing and was used for over 30 years.

During operations at Plum Tree, the sophistication of the drop-model technique dramatically increased. High-resolution video cameras were used for tracking the model, and graphic displays were presented to a

remote pilot control station, including images of the model in flight and the model's location within the range. A high-resolution video image of the model was centrally located in front of a pilot. In addition, digital displays of parameters such as angle of attack, angle of sideslip, altitude, yaw rate, and normal acceleration were also in the pilot's view. The centerpiece of operational capability was a digital ground-based flight control computer programmed with variable research flight control laws and a flight operations computer with telemetry downlinks and uplinks. The drop models continued to be constructed of composite materials, but the instrumentation was upgraded to include three-axis linear accelerometer packages and three-axis angular rate gyro packages.[19]

Operations at Plum Tree lasted about 30 years and included a broad scope of investigations for military aircraft configurations, general-aviation configurations, parawings, gliding parachutes, and reentry vehicles. In the early 1990s, however, several issues regarding environmental protection forced NASA to close its

A fish-eye camera lens captures a scene of a drop model of the X-31 research aircraft being prepared for a drop as part of poststall and spin resistance studies.

research activities at Plum Tree and remove its facilities. NASA assisted in the restoration of the environment to its current status as a restricted access game preserve. Once again, Langley researchers were forced to identify a suitable test site for future drop-model testing. After considerable searching and consideration of several candidate sites, the NASA Wallops Flight Facility was chosen for Langley's drop-model activities.

The most recent drop tests of a military fighter for poststall studies began in 1996 and ended in 2000. This project, which evaluated the spin resistance of a 22-percent-scale model of the U.S. Navy F/A-18E Super Hornet, was the final evolution of drop-model technology for Langley.[20] Launched from a helicopter at an altitude of about 15,000 feet in the vicinity of Wallops, the Super Hornet model weighed about 1,000 pounds. A camera mounted in the cockpit canopy of the model provided an onboard-pilot's view of the flight, while the remote pilot in command sat in an environmentally controlled room at a pilot station with data displays and video images for input cues. The model included extensive instrumentation to provide data to the flight control computer, the data displays for the drop-model cockpit, and postflight data analysis. Recovery of the model at the end of the flight test was again conducted with the deployment of onboard parachutes. The model used a flotation bag after water impact and was retrieved from the Atlantic Ocean by a recovery boat.

An F/A-18E drop model is launched from a NASA helicopter over the NASA Wallops Flight Facility for spin resistance studies. With a weight of 1,000 pounds, the F/A-18E model was the largest drop model ever tested by Langley researchers.

Dryden Unpowered Drop Models

The legendary contributions of the NASA Dryden Flight Research Center in flight-testing full-scale aerospace vehicles, demonstrators, and hardware have been demonstrated. Nearly every major breakthrough in this Nation's aeronautics legacy occurred there, including flights of the X-1, X-15, Shuttle, lifting bodies, X-43, and spacecraft rescue vehicles. The public may be less familiar with the fact that Dryden researchers have conducted a range of important aerospace projects involving the launching of an unpowered test articles from an airborne vehicle. The following is an

Water retrieval of the F/A-18E drop model from the Atlantic Ocean after a drop test. The descent by parachute into the water posed special problems for model designers, instrumentation specialists, and the operational crew.

overview of approaches used at Dryden during tests of subscale unpowered free-flight models to investigate flight behavior of advanced concepts. Results of specific applications will be covered in a later section.

Dryden's primary advocate and highly successful user of free-flight models for low-speed research on advanced aerospace vehicles was the late Robert Dale Reed. An avid model builder, pilot, and enthusiastic researcher, Reed was inspired by his perceived need for a subscale free-flight model demonstrator of the emerging lifting body reentry configuration created by NASA Ames in 1962.[21]

After initial testing of gliders of the Ames M2-F1 vehicle, he progressed into towed models of the configuration released by radio-controlled model tow planes. In the late 1960s, the launching technique for the unpowered models used a powered radio-controlled mother ship, and by 1968, Reed's mother ship had conducted over 120 launches. As will be discussed in a later section, the success of Reed's lifting body model flight work had a dominant effect on subsequent development of the full-scale series of lifting bodies, as well as the Space Shuttle. Dale Reed's innovation and approach to using radio-controlled mother ships for launching drop models of radical configurations has endured to this day as the preferred approach to be used for small-scale free-flight activities at Dryden.

Innovative engineer Dale Reed of the Dryden Flight Research Center used free-flight models to demonstrate concepts for advanced aerospace configurations.

In the early 1970s, Reed's work at Dryden expanded into a series of flight tests of powered and unpowered remotely piloted research vehicles (RPRVs). These activities, which included remote-control evaluations of subscale and full-scale test subjects, used a ground-based cockpit equipped with flight instruments and sensors typical of a representative full-scale airplane. These projects included the Hyper III lifting body and a three-eighths-scale dynamically scaled model of the F-15. The technique used for the F-15 model consisted of air launches of the test article from a B-52 and control by a pilot in a ground cockpit outfitted with a sophisticated control system.[22] The setup featured a digital uplink capability, a ground computer, a television monitor, and a telemetry system. This test technique in 1973 preceded Langley's development of similar sophistication. Initially, the model was recovered on its parachute in flight by helicopter midair snatch, but in later flights, the evaluation pilot landed it on skids. Over 50 flights were made with the F-15 Spin Research Vehicle. Results of Dryden's tests of the 23.5-foot-long model for spin entry research will also be discussed in a later section.

F-15 spin research vehicle mounted to a pylon on its B-52 launch aircraft. The unpowered drop model was the first in a long line of remotely piloted subscale models and was used in conjunction with a piloted simulator at Dryden and a smaller F-15 drop model at Langley to define spin entry maneuvers for the F-15 aircraft.

Langley Powered Free-Flight Models

In addition to the powered free-flight wind tunnel models discussed earlier, NASA has used powered dynamically scaled models at Langley, Ames, and Dryden for investigations of flight dynamics issues for general-aviation aircraft, parawing vehicles, civil transports, and advanced aircraft configurations. In addition to its in-house research and development efforts and cooperative programs with its industry and military partners, NASA has also stimulated the use of powered dynamic models within the university community, greatly increasing the interest and engineering capabilities of students.

At Langley, outdoor powered models have been used in major programs focused on the development of military drones in support of DOD, dynamic stability and control studies of parawing utility vehicles and spacecraft, investigations of general-aviation stall/spin technology, and commercial transport recovery in large-attitude, out-of-control ("upset") situations. Most of these tests used the Plum Tree test area between 1962 and 1992 and simplified versions of the testing techniques previously described. As will be discussed, some of the testing sought to develop relatively inexpensive free-flight techniques that could be applied using modified hobbyist equipment.[23]

NASA conducted research on the spinning charac-teristics of general aviation designs in the mid-1970s with powered radio-controlled dynamically scaled models using modified hobbyist-type gear. These dynamically scaled models were significantly heavier than conventional hobbyist radio-controlled models.

In recent years, Langley has embarked on a sophisticated powered-model technique in support of its Aviation Safety program. The test objectives are to investigate controllability options for multiengine commercial transport-type configurations in upset situations or after airframe/control system damage.

This project, known as the Airborne Subscale Transport Aircraft Research (AirSTAR) program,[24] will be discussed in the next chapter. As part of the activity, Langley researchers have developed a computer-based powered-model capability comparable to the drop-model operations discussed earlier. A special runway for unmanned aerial vehicle (UAV) operations at Wallops is to be used for research operations, and mobile vehicles transported to the site are equipped with sophisticated pilot stations, avionics equipment, and digital computers for simulation of control laws and data acquisition. The commercial transport models are large (with a 7-foot wingspan), dynamically

The NASA AirSTAR program is developing a powered free-flight model technique to study upset recovery and other safety issues for commercial transport configurations. Flight tests of generic powered transport models are underway to minimize risk during studies of a scaled model of the test subject.

scaled, and powered by state-of-the-art model turbine engines. The AirSTAR project is focused on a 5.5-percent dynamically scaled model of the Boeing 757 aircraft, referred to as the Generic Transport Model (GTM). The GTM vehicle weighs about 55 pounds, with approach speeds of about 70 mph. A multiyear developmental process has been underway to develop the pilot skills, instrumentation, and testing required for risk reduction. A phased series of flight projects are being conducted, beginning with relatively low-cost, low-wing-loading models flown near Langley in Smithfield, VA.

Langley Control Line Facility

One of the more unusual free-flight techniques ever developed by Langley resulted from issues regarding the interpretation of wind tunnel free-flight results for vertical take-off and landing (VTOL) models in the early 1960s. The issue arose because many of the VTOL designs exhibited severe stability problems at high angles of attack during the transition from hovering flight to conventional forward flight. Because the airspeed of the Full-Scale Tunnel could not be rapidly increased to simulate a rapid aircraft transition during free-flight model tests, it was argued that, theoretically, an unstable airplane might be able to quickly transition through the unstable range of angle of attack before it could react to the instabilities and experience a loss of control.

Test setup for rapid transition testing of free-flight V/STOL models on the Langley Control Line Facility. Results obtained from the tests generally showed that configurations could minimize their exposure to instabilities during a relatively brief time at high angles of attack, thereby achieving more satisfactory transitions between hovering and forward flight.

Recognizing that this was a fundamental free-flight model issue for this class of aircraft, the Langley staff created an outdoor rapid-transition testing technique.[25] Langley acquired a large crane capable of rapid rotation and equipped it with model power systems and remote-control capability. The crane could be rotated at angular rates up to about 20 revolutions per minute and could accelerate from rest to top speed in about 1/4 revolution. Seated in an enlarged crane cab were the model-controls operator, the safety-cable operator, the model-power operator, and the crane operator. The powered model being tested was mounted about 50 feet from the center of rotation and was restrained by wires from the model to the crane cab to oppose the centrifugal force during flight around the circle. The tests, which resembled the control-line technique traditionally used by model airplane enthusiasts, were limited to studies of the longitudinal motions of the model during a relatively rapid transition process. The testing technique was matured over a few years, and the facility subsequently became known as the Langley Control Line Facility (CLF). In the CLF, transitions could be made much more rapidly than were those in the Full-Scale Tunnel and at rates more representative of those of full-scale VTOL airplanes. Results verified the expected positive result of rapid transitions for tail-sitter, tilt wing, and vectored-thrust V/STOL configurations. The CLF was operational for several tests and was dismantled in 1970.

Ames Powered Free-Flight Models

The NASA Ames Research Center conducted and sponsored outdoor free-flight powered model testing in the 1970s as a result of interest in the oblique wing concept championed by R.T. Jones. The tests included investigations of the flying characteristics of several models of the unorthodox configuration and will be discussed in more detail in the next chapter. The progression of sophistication in these studies started with simple unpowered catapult-launched models at Ames and inspired cooperative testing of powered models at Dryden flight-test sites in the 1970s and piloted flight tests of the AD-1 oblique wing demonstrator aircraft in the 1980s.[26] In the 1990s, Ames and Stanford University collaborated on potential designs for oblique wing supersonic transport designs, which led to flight tests of a 10-foot-span model by Stanford in 1993. In 1994, a 20-foot-span model was designed, constructed, and flown by Stanford at Moffett Field near Ames.

Oblique wing research vehicle flown in cooperative study between Ames and Dryden researchers during the 1970s to explore the aerodynamics and flying characteristics of the configuration. The vehicle was designed to permit its left wing panel to be swept forward 45 degrees. Three research flights in 1976 provided data for the design of the piloted AD-1 research aircraft.

Dryden Powered Free-Flight Models

Dryden's superb flight-test ranges and the expertise of its staff in advanced flight technology have historically been used to evaluate and demonstrate the capabilities of remotely piloted, powered drones, including fighters, atmospheric and environmental samplers, long-duration high-altitude fliers, and solar-powered configurations. To focus on the subject matter of this book, three subscale powered model flight tests are discussed. One project, known as Highly Maneuverable Aircraft Technology (HiMAT), was conducted in the late 1970s using the Dryden RPRV capabilities that had been developed for the previously mentioned B-52 launched unpow-

The X-36 Tailless Fighter Agility Research Aircraft in flight. The sophisticated model demonstrated that a tailless, low-signature configuration could achieve high levels of maneuverability through the use of advanced technology.

ered F-15 model spin investigation.[27] Although a full-scale HiMAT vehicle was not proposed for the program, the RPRV was considered to a subscale version of a larger vehicle. The second subscale Dryden project of relevance was the powered X-36 Tailless Fighter Agility Research Aircraft program, conducted in 1997.[28]

Using a video camera mounted in the nose of the aircraft, the X-36 was remotely controlled by a pilot in a ground station virtual cockpit. A standard fighter-type, head-up display (HUD) and a moving-map representation of the vehicle's position within the range provided situational awareness for the pilot. The third Dryden powered model project to be discussed is the current Dryden flight tests of the X-48B blended wing-body.

Wallops Rocket-Boosted Models

As interest in transonic and supersonic flight increased in the 1940s, the NACA feverishly pursued the development of high-speed wind tunnels, but the presence of tunnel walls contaminated tunnel data until Langley personnel, under the direction of John P. Stack, developed and demonstrated the breakthrough concept of slotted walls. In lieu of a reliable transonic tunnel, researchers turned to the concept of unpowered "free-fall" models to obtain information on aerodynamics, stability, and control at transonic speeds. This effort started in late 1944, when instrumented objects were dropped from NACA research aircraft to acquire transonic data near Langley Field. The pioneering work was accompanied by two other approaches in the attack on transonic phenomena, which consisted of tests of small, instrumented semispan wings mounted in the high-speed flow on the upper wing surfaces of fighter aircraft and a proposal for a high-speed research airplane to be known as the X-1. In 1945, a proposal was made to develop an NACA high-speed test range that would use rocket-assisted launches to explore the transonic and supersonic flight regimes. The facility would ultimately become known as the NACA Wallops Island Flight Test Range. Initially known as the Pilotless Aircraft Research Station, the site launched its first rocket in July 1945.[29] Most of the Wallops tests were designed to explore model characteristics at compressible flight conditions and were therefore subject to the modified scaling procedures mentioned earlier.

F-102 free-flight model is prepared for rocket launched flight at Wallops. Extensive flight testing of hundreds of rocket-boosted models occurred at Wallops during its NACA days. Note that the model has been modified with the wasp-waisted area rule concept conceived by Richard T. Whitcomb of Langley.

From 1945 through 1959, Wallops served as a rocket-model "flying wind tunnel" for researchers at Langley and conducted vital investigations for the Nation's emerging supersonic aircraft, especially the Century series of advanced fighters in the 1950s. Rocket-boosted models were used by Langley's Pilotless Aircraft Research Division (PARD) in flight tests at Wallops to obtain valuable information on aerodynamic drag, stability, and control at transonic conditions.

With the establishment of NASA in 1958 and the immediate focus of the new Agency on higher speeds and spacecraft, Wallops became an independent NASA facility, and its work on aircraft characteristics was replaced by investigations on satellites, sounding rockets, and ballistic missiles. Today, Wallops is recognized as a world leader for launching rockets and satellites, but few remember the key role it played in rocket-boosted free-flight model testing for supersonic aircraft configurations.

CHAPTER 3:

SELECTED APPLICATIONS

The earliest applications of dynamically scaled models at the NACA Langley Memorial Aeronautical Laboratory occurred during research efforts on aircraft spin and spin recovery behavior.[1] The visual results of these investigations provided very effective illustrations of the potentially deadly nature of the spin and the impact of modifications to airplane configurations. Interpretation of the test results was fairly straightforward, and subsequently, the use of free-flight model test techniques was adopted for other technical areas as an integral tool in the research process. Model flight tests offered several important capabilities and advantages beyond those provided by conventional static wind tunnel aerodynamic force tests. The ability to visualize the inherent dynamic stability of an aircraft configuration and its response to atmospheric disturbances as well as to pilot inputs provided a qualitative assessment of flying characteristics and design information on the impact of geometric variables. The relative severity of issues in stability and control was easily assessed, and the ability of modifications to minimize or eliminate problems provided valuable guidance for major design decisions.

For each of the testing techniques previously discussed, initial efforts were directed at correlating results of model and full-scale flight tests to establish confidence in the model test procedures and to develop approaches for interpretation of the test results. Lessons learned in many applications created a fundamental knowledge of potential pitfalls and resulted in modifications to the testing techniques because of factors such as Reynolds number and wind tunnel constraints.

Based on almost 90 years of experience, NASA and its partners have demonstrated the technical advantages of free-flight testing far beyond the visual results. One of the more important advantages of model flight tests is that they provide for continual operations over the flight envelope in terms of flight parameters such as angle of attack and sideslip. In contrast, conventional wind tunnel aerodynamic force tests are constrained to limited combinations of angles of attack and sideslip, and a great deal of experience is required to schedule in advance the potentially critical flight conditions of interest for tunnel tests. Thus, the free-flying model may be thought of as an analog computer with continuous and appropriate aerodynamic inputs (in the absence of Reynolds number or Mach number effects) for an infinite number of flight maneuvers. This capability is extremely valuable in applications in which relatively small changes in angles of attack or sideslip result in large changes in aerodynamic parameters and stability characteristics. When such abrupt changes occur, it can be very difficult to define the specific test conditions that need to be examined in conventional wind tunnel tests.

After a gestation period to establish the credibility of results, NASA's free-flight techniques have been directed toward evaluations and demonstrations of advanced concepts involving unconventional or radical aircraft configurations for which no experience base or design information is available. Another major area of application has been to evaluate the dynamic flight behavior of conventional aircraft operating at flight conditions that involve extremely complex aerodynamic phenomena for which conventional wind tunnel techniques or computational prediction methods have not yet been developed. Finally, many of the free-flight model techniques evolved because it has been virtually impossible to replicate aircraft motions for certain areas of interest within conventional wind tunnels. Such motions include combinations of large-amplitude angular or linear motions.

After the NACA and NASA developed free-flight testing techniques and demonstrated their value in research and development projects, interest in applications quickly developed within the DOD, industry, other Government agencies, and academia. In recognition of the unique national testing capability provided by these techniques, the military services routinely requested high-priority dynamic model tests within NASA facilities to provide early assessments of the flight characteristics of new aircraft designs. In return for predictions of aircraft characteristics and problem-solving projects for the new configurations, DOD provides resources to the NACA and NASA for model fabrication of specific configurations, with the additional benefit to NASA of retaining the models for in-house generic research and development activities. In addition, DOD provides NASA personnel with opportunities to participate in full-scale flight-testing at its test sites and permits researchers to access flight-test data that would otherwise be far beyond NASA's funding capabilities. As a result of these interactions, researchers have participated in almost all DOD aircraft development programs and have established the validity of their free-flight testing techniques while building a vast corporate knowledge of experience for many different aircraft configurations.[2]

NASA provides timely transmission of its results to industry through onsite visits to NASA facilities, technical papers and reports, and industry tours. Industry is encouraged to cooperate in problems of mutual interest in the area of flight dynamics and the use of free-flight models, and this relationship has been especially fruitful for industry during times when corporate research budgets have been tight and futuristic concepts do not meet with near-term managerial interests. Projects to be discussed herein resulted from generic NACA and NASA research, specific DOD requests, and joint projects with industry that involved the use of dynamically scaled free-flight models.

In recent years, NASA has made its historical series of technical reports available for download from its Internet site, and many reports that previously were very limited in distribution, classified, or difficult to obtain are now readily available at *http://ntrs.nasa.gov/search.jsp*. The notes section of the present publication cites many of the papers that provide detailed information on the projects discussed herein, and additional searches for individual authors will provide substantially more material. In addition to having access to the NASA report series, the reader will find supplementary information and discussions of the NASA model results in recent texts on stability and control, such as the excellent book by Malcolm J. Abzug and E. Eugene Larrabee.[3]

The following material presents an overview of the historical applications, lessons learned, and technological impacts of the use of free-flight models for studies of flight dynamics by the NACA and NASA in selected areas through the years. The material has been organized according to the following topics, with illustrative examples of research activities within each:

- Dynamic stability and control.

- Flight at high angles of attack.

- Spinning and spin recovery.

- Spin entry and poststall motions.

- Associated aerodynamic tests.

The discussion begins with a review of NACA and NASA applications of free-flight models to studies of the dynamic stability and control characteristics of a variety of aerospace vehicles across the speed range, from hovering flight at zero airspeed to hypersonic speeds. The research emphasis in such studies was to determine and analyze the dynamic flight motions of configurations within and at the edges of the design envelope. Next, applications of free-flight models in studies of advanced military and civil aircraft designs for critical high-angle-of-attack flight conditions are presented. Historically, flight at high angles of attack with attendant flow separation over the aircraft results in degradation of stability and control, which can lead to inadvertent spins and poststall gyrations. Numerous applications of dynamic models within the area of spinning and spin recovery technology are then discussed. The review then turns to applications in the areas of spin entry and poststall motions that require special outdoor test equipment and ranges. The discussions of applications conclude with overviews of aerodynamic test techniques that are normally performed with the free-flight models to support the analysis of flight dynamics.

CHAPTER 4:

DYNAMIC STABILITY AND CONTROL

Low-Speed Research

Prior to the use of free-flight models, designers relied on qualitative guidelines for aircraft configurations and design details such as the areas required for vertical and horizontal tail surfaces. In some cases, simple theories with broad assumptions were used to predict dynamic motions in response to pilot inputs. The NACA approached flying model experiments with an awareness of potential scale effects that might be caused by the relatively low values of Reynolds number associated with small model tests. After establishing the relative accuracy of results from model tests by correlation with full-scale vehicle experiences, a concentrated effort was undertaken to investigate and demonstrate the effects of configuration variables such as wing planform shape, tail configurations, and other geometric characteristics on the dynamic stability and control characteristics of conventional aircraft designs. By combining free-flight testing with theory, the researchers were able to quantify desirable design features, such as the amount of wing-dihedral angle and the relative size of vertical tail. With these data in hand, methods were developed to theoretically solve equations of motion of aircraft and determine the dynamic stability characteristics such as the frequency of inherent rigid-body oscillations and the damping of those motions.

When Langley began operations of its 12-Foot Free-Flight Tunnel in 1939, high priority was placed on establishing correlation with full-scale flight results. Immediately, requests came from the Army and Navy for correlation of model tests with flight results for the North American BT-9, Brewster XF2A-1, Vought-Sikorsky V-173, Naval Aircraft factory SBN-1, and Vought-Sikorsky XF4U-1. Meanwhile, the NACA used a powered model of the Curtiss P-36 fighter for an in-house detailed calibration of the free-flight process.[1]

The results of the P-36 study were, in general, in fair agreement with airplane flight results, but the dynamic longitudinal stability of the model was found to be greater (more damped) than that of the airplane, and the effectiveness of the model's ailerons was less than that for the airplane. Both discrepancies were attributed to aerodynamic deficiencies of the model caused by the low Reynolds number of the tunnel test and led to one of the first significant modifications to the free-flight technique. The critical lesson learned in

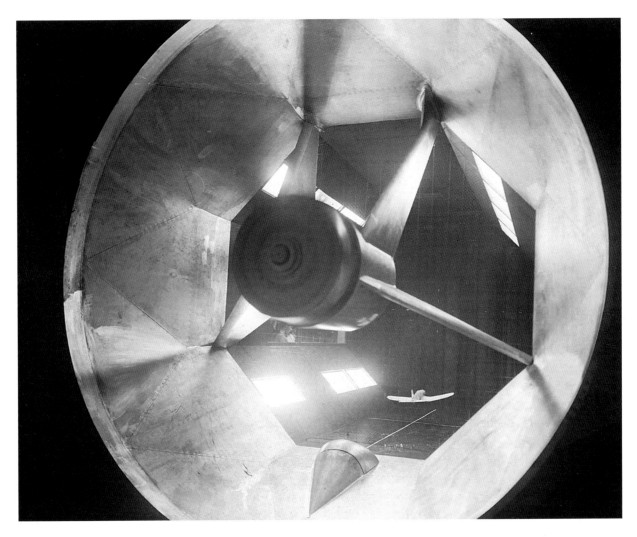

Dynamic stability and control evaluation of the Navy Vought-Sikorsky XF4U-1 Corsair fighter in the Langley 12-Foot Free-Flight Tunnel in September 1940. Early testing in the Free-Flight Tunnel concentrated on correlation of model predictions and full-scale results for dynamic stability and control characteristics.

this early study was that using the specific full-scale P-36 airfoil shape (NACA 2210 airfoil) for the model resulted in poor wing aerodynamic performance at the low Reynolds number of the model flight tests. In particular, a major degradation in lift characteristics was experienced for the 2210 airfoil shape. After this experience, researchers conducted an exhaustive investigation of other airfoils that would have satisfactory performance at low Reynolds numbers. In planning for subsequent tests, the researchers were trained to anticipate the potential existence of scale effects for certain airfoils, even at relatively low angles of attack. As a result of this experience, the wing airfoils of free-flight tunnel models were frequently modified to airfoil shapes that provided better results at low Reynolds number. One specific airfoil substituted for some full-scale wing airfoils in free-flight model testing was the Rhode St. Genese 35 section, which provided a high maximum lift coefficient at low Reynolds number.[2] Even though the modified airfoil affected the fidelity of the predicted characteristics at lower angles of attack, researchers used this approach in attempts to match stall phenomena expected at high angles of attack and full-scale values of Reynolds number.

As early as the 1920s and 1930s, researchers in several wind tunnel and full-scale aircraft flight groups at Langley conducted analytical and experimental investigations to develop design guidelines ensuring satisfactory stability and control behavior.[3] The objective of such studies was to reliably predict the inherent flight characteristics of aircraft as affected by design variables such as the wing-dihedral angle, size and locations of the vertical and horizontal tails, wing planform shape, engine power, mass distribution, and control surface geometry. The staff of the Free-Flight Tunnel joined in these efforts with several studies that correlated the qualitative behavior of free-flight models with analytical predictions of dynamic stability and control characteristics. For example, flight tests of models with excessive wing-dihedral angles dramatically confirmed analytical predictions of undesirable large-amplitude Dutch roll yawing and rolling motions, which were difficult or impossible to control. In other studies, the vertical tail was increased in size, and a marked degradation in the dynamic directional stability was noted, as predicted by the analytical studies. In particular, as the tail size became excessive, the free-flight model exhibited spiral instability, which demanded constant attention and corrective control from the pilot to prevent crashes. Coupled with the results from other facilities and analytical groups, the free-flight results accelerated the maturity of design tools for future aircraft from a qualitative basis to a quantitative methodology, and many of the methods and design data derived from these studies became classic textbook material.[4]

Other factors influencing the dynamic stability of aircraft also received attention during studies in the Free-Flight Tunnel. Fundamental investigations of the effect of mass distribution were conducted as designers began to increase the spanwise distribution of weight through multiengine configurations and flying wing or tailless designs.[5] Again, the research staff very successfully coupled free-flying model tests with analytical predictions of model behavior. In another example of investigations of the effects of physical phenomena on dynamic stability and control, the effects of fuel sloshing in unbaffled fuel cells was explored in the Free-Flight Tunnel.[6] Design trends after World War II had led to enlarged fuselage fuel tanks on high-performance aircraft, and considerable concern had developed over the poten-

NACA free-flight study of the effect of negative wing dihedral. Note the large negative dihedral angle (15 degrees) of the wing. Simple generic models were used in flight studies of the effects of variations in geometric design variables such as dihedral angle, tail size, and center-of-gravity location. Based on the results, design guidelines were formulated.

tial adverse coupling of fuel motions with inherent aircraft dynamic modes of motion to cause degradation of flying qualities. Flight tests of a simple model equipped with liquid-filled spherical fuselage tanks were used to investigate the motion coupling phenomena. The erratic, jerky flight motions of the free-flight model used in the study clearly demonstrated the potential for highly unacceptable flight characteristics that could result if a designer did not appreciate control of internal fuel movements by baffling.

As high-performance aircraft configurations evolved in the early 1950s, the relative length of the fuselage forebody became much longer compared with aircraft of World War II, and in some cases, the cross-sectional shape of the forebody changed from a traditional circular or slab-sided shape to a shape having a relatively flat oval cross section with the major cross-sectional axis horizontal. These design trends resulted in large impacts on aircraft dynamic stability and control. The staff of the Free-Flight Tunnel had encountered an unconventional aerodynamic characteristic related to these features when testing a canard-type configuration in the late 1940s. During those tests, the canard-fuselage combination was directionally unstable at low angles of attack but became stable at high angles of

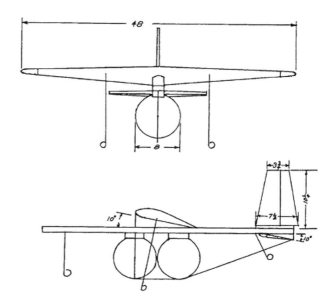

Side and front sketches of the free-flight model used in the fuel sloshing investigation. Spherical fuselage tanks containing various levels of water were used to simulate various arrangements of fuel loading and sloshing characteristics.

attack, at which sidewash from the canard caused a flow reversal on the fuselage so that the combination became directionally stable.[7] As researchers anticipated similar effects for radical oval fuselage cross-sectional shapes, programs were carried out with models in the Free-Flight Tunnel, leading to a compilation of data and design information. As high-performance U.S. military aircraft have evolved, flattened fuselage forebodies have become commonplace for fighters (such as the Northrop F-5 series), and the fundamental understanding derived from the free-flight model research of the 1950s has proven to be valuable guidance for aircraft development programs to this day.

During the final model flight projects in the Free-Flight Tunnel in the mid-1950s, various Langley organizations teamed to quantify the effects of critical aerodynamic dynamic stability parameters on flying characteristics. These efforts included correlation of experimentally determined aerodynamic stability derivatives with theoretical predictions and comparisons of the results of qualitative free-flight tests with theoretical predictions of dynamic stability characteristics. In some cases, rate gyroscopes and servos were used to artificially vary the magnitudes of dynamic aerodynamic stability parameters, such as yawing moment caused by rolling.[8] In these studies, the free-flight model result served as a critical test of the validity of theory.

High-Speed Research: Rocket-Propelled Models

During World War II, tactical maneuvers pushed fighter operations into the relatively unknown realm of aerodynamic compressibility, where new stability and control problems emerged. High-speed dives from high altitudes by fighter pilots had resulted in potentially catastrophic phenomena characterized by excessive stick forces and loss of control in vertical dives. The onset of compressibility effects demanded new methods of analysis, and the NACA responded by developing applications of free-flight models to investigate dynamic stability and control characteristics at transonic and supersonic conditions. As mentioned earlier, in the 1940s, no wind tunnels had been developed that avoided the choking phenomenon that contaminated aerodynamic data measurements near Mach 1, and the NACA developed three alternate approaches to obtain transonic data. Langley used the drop-body, wing-flow, and rocket-model techniques to provide pioneering information on the transonic void.

In 1945, Langley began launching rocket-powered models at its Wallops flight research station to obtain transonic and supersonic information on the performance, stability, and control of generic and specific aircraft configurations. The scope and contributions of the Wallops rocket-powered model research programs for aircraft configurations, missiles, and airframe components covered an astounding number of technical areas, including aerodynamic performance, flutter, stability, control, heat transfer, automatic controls, boundary-layer control, inlet performance, ramjets, and separation behavior. Even a cursory discussion of these results far exceeds the intent of this book, which focuses on dynamic stability and control. The interested reader is, however, referred to the excellent detailed summary of the history and aeronautical contributions of the Wallops Island Flight Test Range by Joseph Shortal.[9]

In addition to providing extensive data via telemetry in several technical disciplines for high-speed conditions, the rocket-powered model tests provided a remarkable amount of information on dynamic stability and control. For example, in just 3 years beginning in 1947, over 386 models were launched at Wallops to evaluate roll control effectiveness at transonic conditions. These tests included generic configurations as well as models with wings representative of the now-famous Douglas D-558-II Skyrocket, Douglas X-3 Stiletto, and Bell X-2 research aircraft. Results on roll control effectiveness at high speeds included pioneering measurements of a severe reduction in rolling effectiveness because of aeroelastic wing effects and the phenomenon in which a complete reversal of intended roll control direction occurred at transonic speeds caused by the aerodynamic behavior of the wing trailing-edge angle.[10] Fundamental studies of dynamic stability and control were also conducted with generic research models to study basic phenomena such as longitudinal trim changes, dynamic longitudinal stability, control-hinge moments, and aerodynamic damping in roll.[11] Studies with models of the D-558-II also detected unexpected coupling of longitudinal and lateral oscillations, a problem that would subsequently prove to be common for configurations with long fuselages and relatively small wings.[12] Similar coupled motions caused great concern in the X-3 and F-100 aircraft development programs and spurred on numerous studies of the phenomenon known as inertial coupling.

Model of the Chance Vought F8U used by the Langley Pilotless Aircraft Research Division at its Wallops flight station in rocket-launched model tests to determine supersonic dynamic stability.

Military requests for rocket-powered model tests in support of high-priority national programs quickly flooded Wallops. The staff made almost continual contributions to the famous aircraft programs of the 1950s. Tests included a series of models of the then-radical XF-92A delta wing airplane.[13] Using NACA-developed instrumentation and telemetry, the test crew at Wallops delivered exceptionally high-quality data on dynamic stability and trim of the XF-92A, resulting in the first-ever predictions of flying qualities based on rocket-model tests. Subsequent results from full-scale airplane flight tests were found to be in good agreement with those obtained from the model studies. In another example, a request from the Navy asked for an evaluation of an unorthodox horizontal-tail control arrangement planned for the Grumman XF10F-1 Jaguar variable-sweep transonic fighter. Rocket-model tests discovered a dynamic instability of the tail surface near Mach 1, as well as a severe directional instability that necessitated the use of ventral fins for the full-scale airplane.[14] More than 20 specific aircraft configurations were evaluated during the Wallops studies, including early models of such well-known aircraft as the Douglas F4D Skyray, the McDonnell F3H Demon, the Convair B-58 Hustler, the North American F-100 Super Sabre, the Chance Vought F8U Crusader, the Convair F-102 Delta Dagger, the Grumman F11F Tiger, and the McDonnell F-4 Phantom II.

The rocket-powered model test technique was not without its detractors. Because each launch resulted in the loss of an expensive model and instrumentation, many of the Langley high-speed wind tunnel staff organizations resented the impact of the rocket model activity on fabrication schedules for model shops and funding for model construction.[15] The growing tension between the groups was finally dissipated in 1958, when the NACA was absorbed by NASA, and the missions of the Wallops and Langley rocket-powered model research organizations were diverted to the space program.

Supersonic Tests at Ames

The quest for high-speed testing techniques at the NACA Ames Laboratory included the development of the Supersonic Free-Flight Tunnel (SSFT) for observing flow fields, shock waves, and dynamic stability exhibited by models launched upstream into a supersonic tunnel flow. Extremely valuable studies of the static and dynamic stability of blunt-nose reentry shapes, including analyses of boundary-layer separation, were conducted in pioneering research efforts. This work included studies of the supersonic dynamic stability characteristics of the Mercury capsule. Spurred on by the experimental observation of nonlinear variations of pitching moment with angle of attack typically exhibited by blunt bodies, Ames researchers contributed a mathematical method for including such nonlinearities in theoretical analyses and predictions of capsule dynamic stability at supersonic speeds.[16] During the X-15 research airplane program, Ames conducted free-flight testing in the supersonic free-flight facility to define stability, control, and flow-field characteristics of the configuration at supersonic speeds.[17]

Unconventional Configurations

Along with the evolving low- and high-speed free-flight analysis capability, progress in other experimental and theoretical procedures rapidly advanced the understanding and prediction of stability and control for conventional configurations. Simultaneously, theoretical methods for predicting the dynamic stability of longitudinal and lateral-directional motions and the effects of control system architecture led to advances in design methods, to the point that textbook design procedures are now adequate for most

Supersonic free-flight test of a model of the X-15 research airplane in the Ames Supersonic Free-Flight Facility. The locations of shock waves emanating from the fuselage, wing, and tail of the model are crisply defined by the shadowgraph technique.

current conventional aircraft development programs. As a result, free-flight dynamic stability testing by the NACA and NASA for conventional aircraft operating in attached-flow conditions within the flight envelope has been very limited. However, a continual introduction of radical aircraft designs for which no experience base exists has challenged designers and focused free-flight applications toward the unconventional configurations. Many of these strange designs were so revolutionary that they generated considerable doubt as to their capability to sustain controlled flight. Arguably, the most valuable contributions of free-flight testing have occurred for these types of vehicles.

The Early Days

Almost immediately after the Langley 12-Foot Free-Flight Tunnel became operational in the 1930s, its unique free-flight capabilities were directed at early assessments of extremely radical, unconventional aircraft designs. By using relatively inexpensive balsa models powered by electric motors, Langley researchers were able to quickly identify major problems, conceive potential solutions, and evaluate the success of modifications.

One of the first radical aircraft configurations to be evaluated by the NACA free-flight technique in 1939 was the Vought-Sikorsky V-173 "Flying Pancake," which had been designed by Charles Zimmerman, the chief advocate and inspiration for the Langley free-flight facilities. Zimmerman, a leading researcher at Langley, had departed in 1937 for Vought, where he proceeded to pursue his concept of a low aspect ratio wing configuration with superior short takeoff and landing capability. The results of the first evaluation flights of a 0.086-scale model of his V-173 design in the Free-Flight Tunnel at Langley were dreadful.[18] In fact, the dynamic stability and controllability of the configuration were so poor that modifications were deemed mandatory before additional testing would be permitted in the tunnel. When the configuration was changed to incorporate outboard horizontal tails, larger vertical tails, and a revised lateral control scheme, the model could be easily flown, and the Navy sponsors of the program were encouraged to proceed to full-scale flight-testing. Flights of the revised model and Zimmerman's other models resulted in continued development of the concept. Highly successful flight-testing of the V-173 during 1942 and 1943 led to development of a second-generation "Zimmer Skimmer," known as the XF5U-1, which arrived too late on the scene at the end of World War II and was canceled in 1947.

In the late 1930s, Jack Northrop's advocacy for flying wing aircraft designs stoked considerable interest within the aeronautical community. However, immediate concern existed over the potential stability and control problems to be encountered with such tailless configurations. Within the NACA, specialists had anticipated potential problems in the areas of yaw control, directional stability,

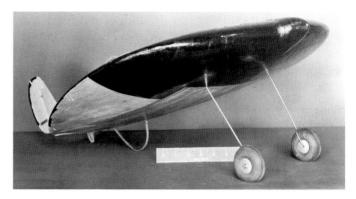

Original unpowered free-flight model configuration of the V-173 "Flying Pancake" as it entered testing in the Langley 12-Foot Free-Flight Tunnel in 1939. Initial test results were very unsatisfactory and required extensive redesign of the tails and control surfaces.

After modifications suggested by powered free-flight model tests at Langley, the V-173 full-scale aircraft was successfully flown and demonstrated outstanding short-field capability.

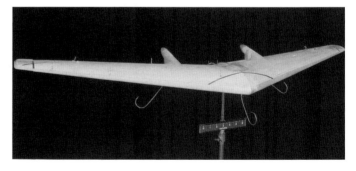

Free-flight model of the Northrop N-9M used in dynamic stability studies in the 12-Foot Free-Flight Tunnel. The test was the first of a series of studies of tailless designs that led to a NACA summary report. Shown here without propellers, the model was also tested for powered conditions.

poststall gyrations, and nose-wheel lift-off speeds for takeoff. In 1941, therefore, the U.S. Army Air Forces requested free-flight tests of the radical Northrop N-9M flying wing configuration in the Free-Flight Tunnel. After a thorough evaluation of the N-9M, a series of flying wing designs was tested, including the Northrop XP-56 "Black Bullet," which was found to have an inadequate-sized vertical tail and poor high-lift capability. During the ensuing decade, a vast amount of NACA generic research was directed at tailless configurations, and the military services requested free-flight model tests of a large number of specific tailless aircraft designs that emerged during that time. For example, the free-flight group at Langley conducted tests of an extremely radical tailless design known as the Kaiser Cargo Wing, a giant flying boat configuration that had been jointly conceived by Howard Hughes and Kaiser. The NACA organizations issued a collaborative summary report on the stability and control characteristics of tailless configurations, including design guidelines for future aircraft.[19]

The introduction of the swept-back wing concept for high-speed drag reduction resulted in a tremendous impact in the field of stability and control. Although they greatly benefitted high subsonic and transonic speed performance, swept wings exhibited very poor low-speed aerodynamic behavior as a result of spanwise flow at moderate angles of attack, resulting in wingtip stall and longitudinal instability in the form of "pitch-up." NACA aerodynamicists at Langley and Ames conducted numerous experimental studies to develop wing modifications and high-lift concepts to minimize or eliminate this problem. In support of this effort, the Langley 12-Foot Free-Flight Tunnel was used to assess the effects of modifications to the leading-edge configuration of the swept wing. Results of these tests were refined and summarized in design methods for calculating dynamic stability and for estimating aerodynamic stability derivatives.[20]

The capture of the German Lippisch DM-1 delta wing glider by advancing American army forces at the end of the war further stimulated a growing U.S. interest in delta wing configurations for future high-speed military aircraft. Full-Scale Tunnel tests of the DM-1 at Langley in 1946 inspired widespread wind tunnel testing and theoretical aerodynamic studies in several organizations. Free-flight model studies of the dynamic stability and control characteristics of delta wing designs were conducted in the Free-Flight Tunnel, including an investigation of the Consolidated Vultee (later Convair) XF-92, which was the first U.S. delta wing fighter. During the precursor testing of the DM-1, NACA researchers had developed an understanding of the beneficial effects on maximum lift caused by vortex flows generated by relatively sharp leading edges and shed over the wing at high angles of attack. In the free-flight tests, these beneficial effects were also noted, in that the model could be flown to very high angles of attack compared with designs with moderately swept wings; however, a violent directional instability was encountered near maximum lift. These dynamic flight results were some of the earliest warnings of potentially degrading vortex flow effects on stability and control at extreme angles of attack. In addition, the model flights indicated that the XF-92's delta wing exhibited pitch-up tendencies at moderate angles of attack, requiring the use of upper-surface wing fences to prevent the instability. The general research conducted on free-flight delta wing models culminated in rocket-powered model-testing of the Convair F-102 configuration at Wallops and low-speed assessments of the design at Langley in the Free-Flight Tunnel.[21]

During World War II, several U.S. and foreign aircraft were designed with unorthodox forward-swept wings, with the objectives of either solving longitudinal center-of-gravity location problems or providing more satisfactory low-speed stall characteristics than those exhibited by aft-swept wings. In contrast to aft-swept wings, which tend to stall at the wingtip, forward-swept wings tend to retain attached flow at the wingtips, thereby maintaining roll control capability of conventional aileron surfaces to high angles of attack. However, the forward-swept wing will tend to stall at the wing root, requiring special consideration in the design process to alleviate undesirable longitudinal stability or trim changes. Researchers at Langley conducted generic and exploratory studies of the stability and control characteristics of this type of configuration in several wind tunnels, beginning in 1946 with an exploratory conventional static force-test study of variable-sweep version of the X-1 research airplane in the Langley 300 mph 7- by 10-Foot Tunnel.[22] In view of the unconventional aerodynamic behavior of forward-swept configurations, the staff of the Free-Flight Tunnel was requested to evaluate the flying characteristics of several forward-swept configurations, including the Convair XB-53 bomber, the Cornelius XFG-1 fuel glider, and the Air Material Command MCD-387-A fighter.[23] Over 30 years later, the forward-swept wing X-29 configuration would undergo free-flight testing in the Full-Scale Tunnel. Results of that project will be discussed in a later section.

Top view of the model of the Convair XB-53 forward-swept wing bomber tested in the Free-Flight Tunnel. The aircraft incorporated a 30-degree swept-forward wing and was tested in several facilities at Langley but did not enter production.

In 1945, military planners became concerned that the emerging Cold War and the strategic threat of the Soviet Union would severely tax the response capability of the U.S. long-range bomber fleet. In particular, the requirement for improving the defensive capability of large aircraft such as the Convair B-36 led to exploration of a parasitic fighter concept in which a small fighter would be carried within the B-36 mother ship and deployed during defensive engagements. After an industry design competition, the Air Force chose

the McDonnell XP-85 design for the mission, and Langley was requested to perform dynamic stability and control assessments of the XP-85 during free-flight tests, which included simulated deployments to and from a trapeze in the bomb bay of the B-36. Results of the model tests indicated the potential for wild gyrations of the aircraft on the trapeze during power-on conditions, and recommendations were made to the Air Force regarding the launch and recovery procedures to avoid stability and control issues.[24] During the first full-scale attempt to engage the trapeze after a successful launch from a converted B-29 bomber at Edwards Air Force Base, the XP-85 hit the trapeze structure, resulting in an emergency landing of the damaged airplane on the desert. The program was canceled in 1949 when air-to-air refueling for fighter escorts became a viable concept.

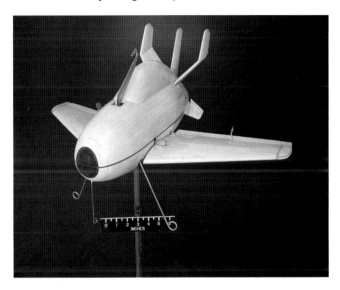

Free-flight model of the McDonnell XP-85 airplane. Note the hinges for the folding wings, which were designed to permit carriage within the bomb bay of the B-36 mother ship, and the fuselage hook used to engage a trapeze in the bay of the B-36.

As the war in the Pacific approached its final stage, interest arose over concepts to extend the range of bomber aircraft and their protective escorts. Industry design teams explored the relative magnitude of aerodynamic drag reduction associated with joining the wingtips of bombers and escorting fighters via tip-hinge

arrangements. The resulting increase in effective aspect ratio for the three aircraft was promising, and the use of a hinged coupling between the aircraft relieved the structural bending loads produced by the aerodynamic and mass properties of the fighter aircraft. However, concern over potential aeroelastic and dynamic stability and control issues that might be experienced in flight represented a major roadblock to the concept. Several generic free-flight model studies were conducted in the Free-Flight Tunnel at Langley to evaluate the effects of geometric details of the wingtip hinge arrangement on dynamic stability

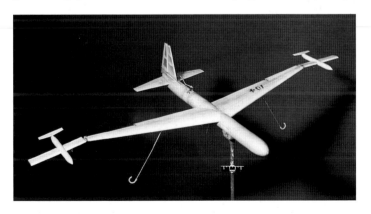

The wingtip-coupling concept was studied in several dynamic stability investigations in the Free-Flight Tunnel. Critical parameters for stability and control such as the fore-and-aft location of the parasite fighters on the bomber's wing tips were identified.

and control, as well as the effectiveness of artificial stabilization systems.[25] The U.S. Air Force sponsored several wingtip-coupling experiments in the early 1950s using B-29 and B-36 mother ships, with F-84 and RF-84 aircraft as parasitic escorts. The last Air Force wingtip-coupling flight experiment in 1956 was known as Project Tom-Tom and used B-36/RF-84F aircraft. The NACA supported this activity with dynamic (not free-flight) hinged-model tests in the Langley 19-Foot Pressure Tunnel. After a series of fatal and near-catastrophic in-flight incidents, the hazards of operational usage of wingtip coupling resulted in a reduction in interest in the concept. Demonstration of the practicality of emerging in-flight refueling techniques for extended range missions was the final blow to the complex tip-coupling approach.

By the time the wind tunnel free-flight model technique was transferred from the Free-Flight Tunnel to the Langley Full-Scale Tunnel in 1959, the interest in model flight tests to determine the dynamic stability and control of unconventional designs was prospering during an age of aeronautical innovation, and a seemingly endless parade of radical designs required analysis.

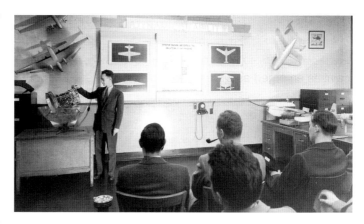

Unconventional configurations are in evidence as John P. Campbell, head of the Free-Flight Tunnel, presents a lecture to visiting aeronautical engineers in 1946. Campbell is describing the operation of the tunnel with a scale model, while the charts and models depict just a few of the radical designs studied in free-flight tests.

Many noted figures in aviation history visited Langley facilities. In this photograph, Gen. James H. "Jimmy" Doolittle completes an inspection of the Free-Flight Tunnel after a lecture by the tunnel staff in 1946.

V/STOL Configurations

Inspired by the demonstrated capabilities of early helicopters, the international aviation community became interested in the late 1940s in vertical take-off and landing (VTOL) aircraft. Around the world, design teams aspired to create VTOL configurations with the capability of helicopter-like hovering flight but with cruise speeds much higher than those of helicopters. The potential advantages of VTOL operations for military applications included a more rapid reaction time for rescue of downed aircrews, dispersal of aircraft operational sites away from easily targeted fixed runways, and the ability to operate on land near the battle area or on small ship-borne landing pads instead of potentially vulnerable aircraft carriers. Interest in VTOL and in vertical/short take-off and landing configurations escalated during the 1950s and persisted through the mid-1960s, with a huge number of radical propulsion/aircraft combinations proposed and evaluated throughout industry, DOD, NACA, and NASA. The multitude of configurations initially included vertical-attitude "tail sitters," followed by an amazing variety of horizontal-attitude designs that used several propulsion concepts to achieve hovering flight and the conversion to conventional forward flight. Tilting wings, rotating shrouded propellers or ducts, vectored nozzles, deflected slipstream flaps, lift fans with vectored louvers, and thrust-augmented cavities emerged as possible propulsion concepts. However, all these concepts were plagued with common issues regarding stability, control, and handling qualities that were experienced in flight.[26]

The extent of research and development for V/STOL aircraft over the years within the United States by Government agencies, industry, and DOD has amounted to a huge investment in resources, including wind tunnel studies, piloted simulator assessments, and flight research evaluations. The details of these activities are far beyond the intent of the present material, which is necessarily limited to highlights involving the use of dynamically scaled free-flight models by the NACA and NASA in coordinated programs with DOD and industry.

The first VTOL nonhelicopter concept to capture the interests of the U.S. military was the vertical-attitude tail-sitter concept. The postwar discovery that German Focke-Wulf engineers had studied a tail-sitter concept known as the "Triebflugel" (thrust wing) vertical-attitude fighter in 1944 was a major stimulus for this interest. The Focke-Wulf paper study featured three unswept wing panels that were powered by wingtip-mounted ramjets and rotated around the fuselage at incidence to provide lift for vertical takeoff. In 1947, the Air Force and Navy initiated an activity known as Project Hummingbird that requested design approaches for vertical take-off aircraft. At Langley, discussions with Navy managers led to exploratory evaluations of simplified free-flight, tail-sitter models to evaluate stability and control during hovering flight in 1949. Conducted in a large open area within a building, powered-model testing enabled researchers to assess the dynamic stability and control of such configurations.[27] The test results provided extremely valuable information on the relative severity of unstable oscillations encountered during hovering flight. The instabilities in roll and pitch were caused by aerodynamic interactions of the propeller during forward or sideward translation, but the period of the growing oscillations was sufficiently long to permit relatively easy control. The model flight tests also provided guidance regarding the level of control power required for satisfactory maneuvering during hovering flight.

Navy interest in the tail-sitter concept led to contracts for the development of the Consolidated Vultee (later Convair) XFY-1 "Pogo" and the Lockheed XFV-1 "Salmon" tail-sitter aircraft in 1951. The Navy requested that Langley conduct

The first NACA exploratory tail-sitter VTOL free-flight model. Powered by an electric motor, the model was flown in hovering flight in a large open building for evaluations of its inherent dynamic stability and the controllability of instabilities.

dynamic stability and control investigations of both configurations using its free-flight model test techniques. In 1952, hovering flights of the Pogo were conducted within a large area in the return passage of the Langley Full-Scale Tunnel, followed by transition flights from hovering to forward flight in the tunnel test section during a brief break in the tunnel's busy test schedule.[28] Observed by Convair personnel (including the XFY-1 test pilot), the flight tests provided encouragement and confidence to the visitors and the Navy. An important out-

put from the transition tests in the tunnel was the recognition that instabilities encountered during the relatively long time required to increase the tunnel speed during the transition would probably be minimized during more realistic unconstrained transition maneuvers, which would require much less time. Langley researchers conducted additional rapid-transition free-flight tests at the Langley Control Line Facility (CLF). As expected, the rapidity of the transition maneuver made the pilot's task much easier at the control-line facility.[29] As a result of experiences with the XFY-1 model, other NASA VTOL free-flight tests were conducted using the CLF, including the X-13 tail-sitter, the Hawker P.1127 vectored-thrust configuration, and the VZ-2 tilt wing design.[30]

NACA's support for the Lockheed XFV-1 tail-sitter program included free-flight model testing in both the Langley Full-Scale Tunnel and, for the first time, the Ames 40- by 80-Foot Tunnel.[31] The tests at Ames required the development of a testing technique for the closed-throat test section of the 40- by 80-Foot Tunnel and the use of a 9.5-foot-long, 250-pound free-flight model.[32] In addition to providing pilot inputs, the control system included artificial damping in pitch, roll, and yaw. The model was powered by electric motors and included a trailing flight cable. The hovering flight evaluations were conducted in the return passage of the tunnel near the contraction cone with three safety cables attached to the wingtips and nose of the model, and personnel were stationed to disperse the lines via a high-speed winch.

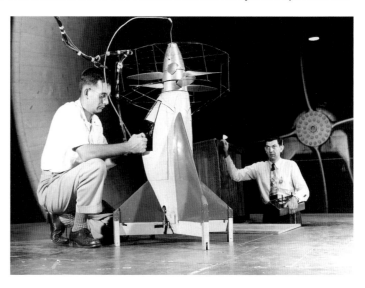

Flight studies of the stability and control characteristics of the Convair XFY-1 were conducted in the Langley Full-Scale Tunnel. Note the flexible flight cable, the one-eighth-inch steel safety cable, and the wire propeller guard.

Test setup for hovering flights of a large free-flight model of the Lockheed XFV-1 in the return passage of the Ames 40- by 80-Foot Tunnel.

In contrast to tests at Langley, the Ames setup used only a single pilot and a model-power operator. For the transition flights, the model was flown in the test section, and the test crew was stationed outside the wind tunnel, with the exception of the two-member tether crew. The pilot's field of vision was somewhat restricted, because he was outside the tunnel and had to observe the model motions through test section windows. This unique investigation provided information on the necessary flight control system characteristics and, along with the testing at Langley, provided confidence to the Navy sponsors that the configuration could be controlled in VTOL operations. To the author's knowledge, the XFV-1 free-flight tests have been the only free-flight tests conducted in the Ames tunnel. Because the Ames 40- by 80-Foot Tunnel provided higher test speed capabilities and was the Nation's premier subsonic wind tunnel for full-scale configurations, it was used as the primary NACA and NASA large low-speed test facility for conventional testing, whereas the Langley Full-Scale Tunnel was the primary facility for free-flight testing of subscale models.

Setup for free-flight transition tests of XFV-1 model in the Ames 40- by 80-Foot Tunnel.

Although the tail-sitter VTOL concept was successfully demonstrated during free-flight model studies and flight demonstrations of full-scale aircraft such as the Convair XFY-1 and Ryan X-13, a multitude of operational issues were encountered that limited the operational feasibility of the concept, and the engineering community abandoned the tail-sitter approach. Chief among these problems was the unsatisfactory situation of having the pilot positioned in a reclined attitude during vertical take-offs and landings, resulting in impaired vision and lack of visual cues for landing. This visual shortcoming proved to be the most serious inherent deficiency in the tail-sitter concept and led to widespread disinterest during the 1960s, when new horizontal-attitude concepts appeared as candidates for V/STOL missions.

In the 1970s, the emergence of the General Dynamics YF-16 and Northrop YF-17 lightweight fighter prototypes inspired Langley researchers to reexamine the tail-sitter concept.[33] As originally conceived, these highly

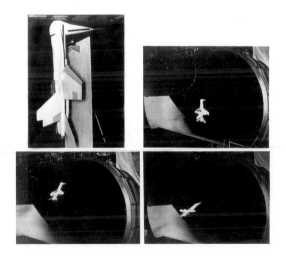

Langley's version of a hypothetical YF-17 "Woodpecker" VTOL fighter. Based on an installed thrust-to-weight ratio of over 1, a fly-by-wire flight control system, and a tiltable cockpit, the design would take off and land on a vertical platform using a hook-and-bar arrangement. It could also conduct take-offs and landings from runways during conventional operations. The sequence shows conversion from hovering flight to conventional flight in the Full-Scale Tunnel.

maneuverable prototypes had thrust-to-weight ratios greater than 1 and included fly-by-wire flight control systems without traditional control linkages and hardware. NASA researchers theorized that the use of fly-by-wire might permit the possibility of maintaining the pilot in a conventional horizontal attitude using a rotatable cockpit arrangement, and they evaluated the dynamic stability and control characteristics of research versions of each of the lightweight fighters, known as VTOL "Woodpecker" concepts.

Highly successful hovering and transition flights to and from conventional wing-borne flight were conducted for the modified YF-16 and YF-17 models in the Full-Scale Tunnel, and results were briefed to industry and DOD. Grumman, which had been following the NASA experiments, pursued a refined version of the basic concept in a project known as the "Nutcracker," which was to be a 21,500-pound, turbofan-powered Navy antisubmarine aircraft. Grumman built and flight-tested a radio-controlled model and successfully demonstrated the docking procedure for take-off and landing.[34]

NASA free-flight model of the Hawker P.1127 VTOL airplane hovers in the return passage of the Full-Scale Tunnel. Free-flight tests and transonic propulsion integration tests in the Langley 16-Foot Transonic Tunnel were key contributions to the successful development of the airplane.

Without doubt, the most successful NASA application of free-flight models for VTOL research was in support of the British P.1127 vectored-thrust fighter program. As the British Hawker Aircraft Company matured its design of the revolutionary P.1127 in the late 1950s, Langley's management became staunch supporters of the activity and directed that tests in the 16-Foot Transonic Tunnel and free-flight research activities in the Full-Scale Tunnel be used for further development work.[35] In response to the directive, a one-sixth-scale free-flight model was flown in the Full-Scale Tunnel to examine the hovering and transition behavior of the design.[36] Results of the free-flight tests were very impressive and were witnessed by Hawker staff members, including the test pilot slated to conduct the first transition flights. The NASA researchers regarded the P.1127 model as the most docile V/STOL configuration ever flown during their extensive experiences with free-flight VTOL designs. As was the case for many free-flight model projects, the motion-picture segments showing successful transitions from hovering to conventional flight in the Full-Scale Tunnel and on the control-line facility were a powerful influence in convincing critics that the concept was feasible. In this case, the model flight demonstrations helped sway a doubtful British government to fund the project. Ultimately, refined versions of the P.1127 design were developed into today's Harrier and AV-8 fighter-attack aircraft.

The NACA and NASA also conducted extensive free-flight model research on tilt wing aircraft for V/STOL missions. In the early 1950s, several generic free-flight propeller-powered models were flown to evaluate some of the stability and control issues that were anticipated to limit the application of the concept.[37] The fundamental principle used by this concept to convert from hovering to forward flight involves reorienting the wing from a vertical position for takeoff to a conventional position for forward flight. However, this simple conversion of the wing angle relative to the fuselage brings major challenges. For example, the wing experiences large changes in angle of attack relative to the flight path during the transition, and areas of wing stall may be encountered during the maneuver. Wing stall can result in wing-dropping, wallowing motion, and uncommanded transient maneuvers. Therefore, the wing must be carefully designed to minimize or eliminate flow separation that could result in degraded or unsatisfactory stability and control characteristics. Extensive wind tunnel and flight research on many generic models, as well as the Hiller X-18, Vertol VZ-2, and Ling-Temco-Vought XC-142A tilt wing configurations at Langley, included a series of free-flight model tests in the Full-Scale Tunnel.[38]

Model of the VZ-2 tilt wing VTOL research airplane during transition flight tests in the Full-Scale Tunnel. Although the tunnel's airspeed acceleration was too slow to permit rapid transitions, assessments of wing stall and dynamic stability characteristics could be made for fixed airspeeds in a controlled manner.

Coordinated closely with full-scale flight tests, the model testing focused on providing early information on dynamic stability and the adequacy of control power in hovering and transition flight for the configurations. However, all projects quickly encountered the primary problem area of wing stall, especially in reduced-power descending flight maneuvers. Tilt wing aircraft depend on the high-energy slipstream of propellers to prevent local wing stall by reducing the effective angle of attack across the wingspan. For reduced-power conditions, which are required for steep descents to accomplish short-field missions, the energy of the slipstream is severely reduced, and wing stall is experienced. Large uncontrolled dynamic motions may be exhibited by the configuration for such conditions, and the undesirable motions can limit the descent capability (or safety) of the airplane. Flying model tests provided valuable information on the acceptability of uncontrolled motions, such as wing dropping and lateral-directional wallowing during descent, and the test technique was used to evaluate the effectiveness of aircraft modifications such as wing flaps or slats, which were ultimately adapted by full-scale aircraft, such as the XC-142A.

To simulate the critical state of descending flight in the fixed horizontal airflow of the Full-Scale Tunnel, Langley developed a modification to the free-flight test technique. An auxiliary compressed-air thrust tube was placed in the rear fuselage of the test model, and following a series of carefully calibrated conventional force tests of the model at various propeller-powered conditions, the auxiliary thrust was used in part to balance drag and permit simulation of reduced propeller power for descent. Although not fully correct from a technical perspective, this pseudo-descent simulation permitted the test team to obtain first-order critical information prior to flight tests of the full-scale aircraft.

As the 1960s drew to a close, the worldwide engineering community began to appreciate that the weight and complexity required for VTOL missions presented significant penalties in aircraft design. It therefore turned its attention to the possibility of providing less demanding short take-off and landing capability with fewer penalties, particularly for large military transport aircraft. Langley researchers had begun to explore methods of using propeller or jet exhaust flows to induce additional lift on wing surfaces in the 1950s, and although the magnitude of lift augmentation was relatively high, practical propulsion limitations stymied the application of many concepts.

A particularly promising concept known as the externally blown flap (EBF) used the redirected jet engine exhausts from conventional pod-mounted engines to induce additional circulation lift at low speeds for take-off and landing.[39] However, the relatively hot exhaust temperatures of turbojets of the 1950s were much too high for structural integrity and feasible applications. Nonetheless, Langley continued to explore and mature such ideas, known as "powered-lift" concepts. These research studies embodied conventional-powered model tests in several wind tunnels, including free-flight investigations of the dynamic stability and control of multiengine EBF configurations, with emphasis on providing satisfactory lateral control and lateral-directional trim after the failure of an engine. Other powered-lift concepts were also explored, including the upper-surface-blowing (USB) configuration, in which the engine exhaust is directed over the upper surface of the wing to induce additional circulation and lift.[40] Advantages of this approach included potential noise shielding and flow-turning efficiency.

While Langley continued its fundamental research on EBF and USB configurations, in the early 1970s, an enabling technology leap occurred with the introduction of turbofan engines, which inherently produce relatively cool exhaust fan flow. The turbofan was the perfect match for these STOL concepts, and industry's awareness and participation in the basic NASA research program matured the state of the art for design data for powered-lift aircraft. The free-flight model results, coupled with NASA piloted simulator

John P. Campbell, left, inventor of the externally blown flap (EBF) concept, and Gerald G. Kayten, of NASA Headquarters, pose with a generic free-flight model of an STOL configuration at the Full-Scale Tunnel. Large slotted trailing-edge flaps were used to deflect the exhaust flows of turbofan engines to induce extremely high lift at low speeds. Note the nose-down tilt of the engine nacelles and the high T-tail used to avoid the large downwash variations behind the wing in STOL conditions.

studies of full-scale aircraft STOL missions, helped provide the fundamental knowledge and data required to reduce risk in development programs. Ultimately applied to the McDonnell-Douglas YC-15 and Boeing YC-14 prototype transports in the 1970s and today's Boeing C-17, the EBF and USB concepts were the result of over 30 years of NASA research and development, including many valuable studies of free-flight models in the Full-Scale Tunnel.

Parawing Configurations

Langley's Francis Rogallo is widely recognized as the father of the flexible wing concept known as the parawing. Rogallo conceived the idea of a flexible, diamond-shaped wing attached to rigid wing-forming members in 1947. His efforts to demonstrate the gliding potential of unpowered parawing configurations were followed by a growing interest in industry for broad applications to utility vehicles and cargo delivery concepts. When the space age dawned, NASA began to explore various candidate approaches for the recovery of its space capsules after exploratory space missions. A special interest was the possibility of extending the landing footprint to the point that recovery could occur on land or runways, thereby avoiding the

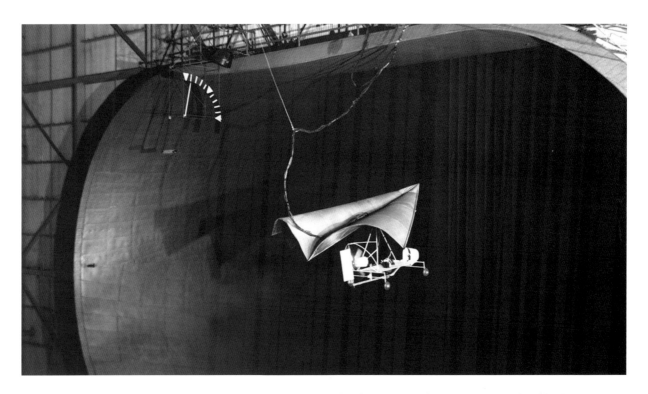

Free-flight model of parawing utility vehicle during tests in the Langley Full-Scale Tunnel. Control was accomplished by tilting the parawing relative to the pilot's platform. After large adverse control effects were discovered using the technique, researchers conceived new roll control concepts using hinged keel members.

complexity and cost of ballistic water landings, which involved recovery ships and other complications. The search for recovery concepts that provided a degree of modification to the ballistic recovery flight trajectory included serious considerations of parawing/capsule applications. At the same time, free-flight studies were conducted to explore the use of packaged, deployable parawings for the recovery of rocket boosters, such as the Saturn rocket.

Before any of the foregoing parawing applications could be realized, significant research was required on the dynamic stability and control characteristics of the unconventional configurations resulting from parawing-vehicle combinations. In addition to traditional stability and control issues, such as how to provide sufficient levels of control without excessive adverse effects, researchers set out to explore problems associated with the deployment of flexible wings from other vehicles during flight.[41] Free-flight model testing of the dynamic stability of parawing vehicles began with a series of studies of powered parawing utility vehicles in 1961.[42] One of the major issues to be addressed in the design of parawing/vehicle combinations was the adequacy of longitudinal and lateral-directional control, because the relatively high position of the parawing relative to the center of gravity of the vehicle could create unconventional responses to inputs by the pilot.

During flight tests of dynamic models, it was found that shifting the center of gravity of the vehicle fore and aft for pitch control and side to side for roll control was satisfactory for some configurations. This control concept was implemented by banking the wing relative to the vehicle for roll control and by pitching the wing for longitudinal control. However, other configurations using the same control technique exhibited marginal or unsatisfactory levels of control. In fact, some of the configurations would roll in a direction opposite to the pilot's input because of excessive adverse yaw. Langley researchers conceived and demonstrated a revised lateral control concept, in which wing bank was still used for roll control, but the aft sections of the rigid leading edge of the parawing were hinged to reduce large hinge moments produced by the original wing-bank control system. When modified with the revised control concept, the free-flight models all exhibited satisfactory characteristics.[43]

In the early 1960s, parawing recovery of large rocket boosters in the class of the Saturn rocket was explored by wind tunnel testing and outdoor drop-model studies at Langley's Plum Tree test range.[44] The concept used a folded rigid-member parawing that was stowed and deployed from the side of the booster after launch and booster burnout during an essentially vertical flight path at subsonic speeds. After deployment, a small drogue parachute was used to maintain separation between the booster and the parawing until the parawing was fully deployed and producing lift for the recovery pullout maneuver. Suspension lines at the front and rear of the parawing were used for control inputs during the recovery flight. Although simple in concept, this recovery procedure had many potential pitfalls that were identified and eventually solved

during the test program. Critical design parameters such as the length of the suspension lines, the spacing between the side of the booster and the inflating parawing, and the opening shock loads were all assessed using the drop-model testing technique, in which the booster and stowed parawing were released from a helicopter at altitudes of about 3,000 feet. Once a satisfactory deployment was obtained and the parawing/booster combination entered a glide, stability and control problems were noted, including an abrupt stall that sometimes led to the test article pitching down and tumbling with the booster falling on top of the parawing. The free-flight model also exhibited relatively small constant-amplitude rolling and yawing motions.

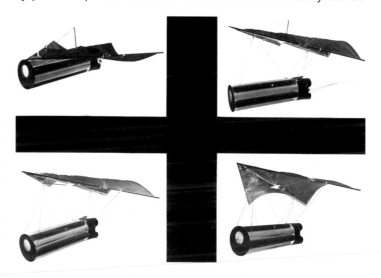

Sequence showing deployment process for a free-flight model parawing recovery system for the Saturn rocket booster. Dynamic stability and control challenges were met after considerable research with a large free-flight model deployed from a helicopter.

Many organizations at Langley were also involved in wind tunnel testing of parawing/capsule combinations to assess aerodynamic characteristics for possible applications as gliding recovery systems. These activities included free-flight model testing in the Full-Scale Tunnel and outdoor drop-model testing.[45] Once again, the parawing deployment process and an evaluation of the dynamic stability and control of the parawing/capsule arrangement was the major thrust of the research efforts. The drop-model used a telescoping rigid parawing in which the leading edges could be retracted into a small spanwise dimension and attached near the apex of a blunted cone suspended below the parawing. The lengths of the suspension lines were varied to shift the center of gravity of the vehicle fore and aft for pitch control and side to side for roll or lateral control. A drogue parachute was necessary to extract, open, and separate the parawing from the capsule. Dropped from a helicopter at near-zero airspeed and an altitude of approximately 3,000 feet, the 85-pound combination began the parawing deployment process at about 2,000 feet. After a series of exploratory test flights established the critical design parameters, results indicated that the free-flight model was stable and could be controlled during gliding flight by shifting the center of gravity. The deployment of the parawing, however, required a carefully controlled sequence that could not occur too quickly. In particular, the parawing had to be slowly rotated to a lifting condition, or a tumbling motion ensued.

In addition to the foregoing studies, many more free-flight model investigations of parawing configurations were conducted at Langley with a variety of applications, including landing aids for high-speed aircraft and reentry vehicles.[46]

Variable Geometry

Spurred on by postwar interests in the variable wing-sweep concept as a means to optimize mission performance at both low and high speeds, the NACA at Langley initiated a broad research program to identify the potential benefits and problems associated with the concept.[47] Early exploratory wind tunnel studies of a modified X-1 model have already been mentioned. The disappointing experiences of the Bell X-5 research aircraft, which used a single wing pivot to achieve variable sweep in the early 1950s, had clearly identified the unacceptable weight penalties associated with the concept of translating the wing along the fuselage centerline to maintain satisfactory levels of longitudinal stability while the wing-sweep angle was varied from forward to aft sweep. After the X-5 experience, military interest in variable sweep quickly diminished, while the NACA continued to explore alternate concepts that might permit variations in wing sweep without moving the pivot location and without serious degradation in longitudinal stability and control.

After years of intense research and wind tunnel testing, Langley researchers conceived a promising concept known as the outboard pivot.[48] The basic principle of the NASA solution was to pivot the movable wing panels at two outboard pivot locations on a fixed inner wing and share the lift between the fixed portion of the wing and the movable outer wing panel, thereby minimizing the longitudinal movement of the aerodynamic center of lift for various flight speeds. As the concept was matured in configuration studies and supporting tests, refined designs were continually submitted to intense evaluations in tunnels across the speed range from supersonic cruise conditions to subsonic takeoff and landing.

Free-flight model of outboard pivot variable-sweep model used in research studies at Langley. Conceived by researchers at the Langley 7- by 10-Foot High-Speed Tunnel, this configuration was tested across the speed range, including low-speed flight tests in the Full-Scale Tunnel.

The use of dynamically scaled free-flight models to evaluate the stability and control characteristics of variable-sweep configurations was an ideal application of the testing technique. Because variable-sweep designs are capable of an infinite number of wing-sweep angles between the forward and aft positions, the number of conventional wind tunnel force tests required to document stability and control variations with wing sweep for every sweep angle could become unacceptable. In contrast, a free-flight model with continually variable wing sweep angles could be used to quickly examine qualitative characteristics as its geometry changed, leading to rapid identification of problems. Free-flight model demonstrations of a configuration based on a proposed Navy combat air patrol (CAP) mission in the Full-Scale Tunnel provided a convincing demonstration that the outboard pivot was ready for applications. This fundamental research program provided key guidance for the design of fighters and bombers for the U.S. military, including the General Dynamics F-111 and the Rockwell B-1 for the Air Force, and the Grumman F-14 for the Navy. During the early configuration development of these variable-sweep aircraft, Langley conducted free-flight, spin tunnel, and outdoor drop-model tests of dynamically scaled models of each configuration at the request of DOD to evaluate dynamic stability and control, spin and recovery, and spin resistance characteristics.

The NACA and NASA have explored other approaches to providing the aerodynamic benefits of variable wing sweep. The oblique wing concept (sometimes referred to as the "switchblade wing" or "skewed wing") had originated in the German design studies of the Blom & Voss P202 jet aircraft during World War II and was pursued at Langley by Robert T. Jones. Oblique wing designs use a single-pivot, all-moving wing to achieve variable sweep in an asymmetrical fashion. The wing is positioned in the conventional unswept position for take-off and landings, and it is rotated about its single pivot point for high-speed flight. As part of a general research effort that included theoretical aerodynamic studies and conventional wind tunnel tests, a free-flight investigation of the dynamic stability and control of a simplified model was conducted in the Free-Flight Tunnel in 1946.[49] This research

Flight test of a skewed wing free-flight model with the wing swept 40 degrees in the Free-Flight Tunnel in 1946. R.T. Jones and John P. Campbell initiated these exploratory tests to assess the flight behavior of the asymmetric configuration and were the first flight tests of the skewed wing concept in the U.S. aeronautical community.

on the asymmetric swept wing actually predated NACA wind tunnel research on symmetrical variable sweep concepts with the X-1 model.[50] The test objectives were to determine whether such a radical aircraft configuration would exhibit satisfactory stability characteristics and remain controllable in the swept wing asymmetric state at low-speed flight conditions. The results of the tests, which were the first U.S. flight studies of oblique wings, showed that the wing could be swept as much as 40 degrees without significant degradation in behavior. However, when the sweep angle was increased to 60 degrees, an unacceptable longitudinal trim change was experienced, and a severe reduction in lateral control occurred at moderate and high angles of attack. Nonetheless, the results obtained with the simple free-flight model provided optimism that the unconventional oblique wing concept might be adaptable.

R.T. Jones transferred to the NACA Ames Aeronautical Laboratory in 1947 and continued his brilliant career, which included a continuing interest in the application of oblique wing technology. In the early 1970s, the scope of NASA studies on potential civil supersonic transport configurations included an effort by an Ames team headed by Jones that examined a possible oblique wing version of the Supersonic Transport (SST). Although wind tunnel testing was conducted at Ames, the demise and cancellation of the American SST program in the early 1970s terminated this activity. Wind tunnel and computational studies of oblique wing designs continued at Ames throughout the 1970s for subsonic, transonic, and supersonic flight applications.[51] Jones participated in flight tests of several oblique wing, radio-controlled models, and a joint

Ames-Dryden project was initiated to use the remotely piloted Oblique Wing Research Aircraft (OWRA) for studies of the aerodynamic characteristics and control requirements to achieve satisfactory handling qualities. The wing of the OWRA could be skewed 45 degrees (left wing forward), and the vehicle was tested in the Ames 40- by 80-Foot Tunnel during the developmental program. A successful flight program at Dryden provided critical information for a follow-on piloted demonstrator.

The AD-1 oblique wing demonstrator in flight at the Dryden Flight Research Center in 1981. The aircraft completed 79 research missions.

Growing interest in the oblique wing and the success of the OWRA remotely piloted vehicle project led to the design and low-speed flight demonstrations of a full-scale research aircraft known as the AD-1 in the late 1970s. Designed as a low-cost demonstrator, the radical AD-1 proved to be a showstopper during air shows and generated considerable public interest. The flight characteristics of the AD-1 were satisfactory for wing-sweep angles of less than about 45 degrees, but the handling qualities degraded for higher values of sweep, in agreement with the earlier Langley exploratory free-flight model study.

After his retirement from NASA in 1981, Jones continued his interest in supersonic oblique wing transport configurations. When the NASA High Speed Research program to develop technologies necessary for a viable Supersonic Transport began in the 1990s, several industry teams revisited the oblique wing. Ames sponsored free-flight, radio-controlled model

Stanford University supported the NASA Ames Research Center with radio-controlled, free-flight model testing of several oblique wing models. Here, a 20-foot-span oblique flying wing model powered by two ducted fans is undergoing flight evaluations.

studies of oblique wing configurations at Stanford University in the early 1990s.[52] The first Stanford model of a radical oblique flying wing (OFW) design was a propeller-powered, 10-foot span with wing sweep capability for angles between 25 and 65 degrees. A second free-flight model had a 20-foot span and was designed to be a 0.05-scale model of a 400-passenger aircraft. Powered by two ducted fans, the model instrumentation included a three-axis rate gyro, angle-of-attack and sideslip vanes, and a turbine airspeed indicator. After a preflight developmental process mounted to a three-degree-of-freedom rig atop an automobile, the model was flown in 1994 with a sweep-angle variation from 35 to 50 degrees.

As a result of free-flight model contributions from Langley, Ames, Dryden, and academia, major issues regarding potential dynamic stability and control problems for oblique wing configurations have been addressed for low-speed conditions. Unfortunately, funding for transonic and supersonic model flight studies has not been forthcoming, and high-speed studies have not yet been accomplished. A 2006 oblique flying wing contract to Northrop Grumman by the Defense Advanced Research Projects Agency (DARPA) was initiated to mature the technology to achieve high cruise efficiency, long endurance, and extended low-speed loiter capability. The program objectives included the development and flight of a small-scale supersonic technology demonstrator X-plane known as Switchblade, and wind tunnel testing was underway in 2007. In October 2008, DARPA canceled the project after the preliminary design effort.

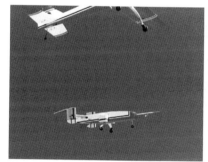  

Launch and deployment of a foldout wing demonstrator model at the NASA Dryden Flight Research Center using an unpowered test model and the mother ship technique.

Other examples of free-flight models used to explore the effects of variable geometry on flight characteristics include the use of foldout wings and lifting surfaces and extensible stabilizing surfaces. For example, in 2001, Dryden researchers used Dale Reed's mother ship technique to launch an unpowered research model that demonstrated a deployable, inflatable wing.[53] Known as the I2000 configuration, the instrumented research model was air-launched from a larger radio-controlled model at an altitude of about 1,000 feet, and its deployable wings were inflated by an onboard nitrogen system. Successful flights were conducted with the model demonstrating satisfactory stability during the wing deployment from the stowed position on three-drop operations, and parameter estimation procedures were used to extract aerodynamic properties of the model.

Arguably, the most challenging NASA application of free-flying models to variable geometry requirements has been a result of interplanetary exploration interests.[54] The age-old dream of using remotely piloted aircraft to explore Mars continually resurfaces as an attractive alternative to land rovers or orbiting satellites. An aircraft flying about 1 mile above the Martian surface could obtain high-resolution mapping over a 400-mile distance with navigation to specific areas. In addition to the challenges in developing this capability within the areas of propulsion, structures, and aerodynamics, an approach must be devised to design and construct a stowable aircraft aboard an interplanetary rocket and deploy it into conventional flight. Over the past 40 years, NASA, its industry partners, and universities have addressed a majority of these concerns through mission studies, wind tunnel tests, propulsion and propulsion integration studies, and vehicle design studies. The concept calls for delivery of the folded aircraft to Mars within a protective aeroshell aboard a spacecraft. After the aeroshell separates from the carrier, it enters the Martian atmosphere, where it begins to decelerate, and a parachute deployment provides additional deceleration. After a heat shield on the aeroshell is released, the stowed aircraft is released and unfolded, the pullout maneuver to conventional flight is completed, and the vehicle begins its exploration mission.

In the late 1990s, NASA invited proposals for Mars exploration from the scientific community, including Langley, which had organized disciplinary experts into a design team for a Mars airplane. By 2002, specialized wind tunnel tests to develop a deployable configuration were underway, including dynamic tests to investigate design approaches for the deployment. In mid-2002, development of a proposal known as Aerial Regional-Scale Environmental Survey (ARES) of Mars was submitted to NASA Headquarters for competitive evaluations for Mars exploratory missions.[55] Langley had

NASA–industry test team poses with the Mars airplane High-Altitude Deployment Demonstrator model before its flight.

teamed with Aurora Applied Sciences, Corp., of Manassas, VA, for flight tests of candidate aircraft configurations. One aircraft, known as the High-Altitude Deployment Demonstrator 1 (HADD1), was used in critical high-altitude deployment tests. Carried aloft by a high-altitude helium balloon to about 100,000 feet over Oregon, the 10-foot-span HADD1 was released from the balloon and unfolded, and it completed a 90-minute, preprogrammed autonomous flight. After the aircraft completed a successful flight, control transferred to a human pilot for a safe landing. A second aircraft, known as HADD2, was a full-scale version of the Mars aircraft with a 20-foot span, and it was delivered to Langley in 2006 to be prepared for instrumentation. After a review of all proposals for the Mars exploration mission, NASA Headquarters selected proposals for satellite orbiters, and the Mars airplane activity at Langley was terminated.

Space Capsules

The selection of blunt capsule designs for the Mercury, Gemini, and Apollo programs resulted in numerous investigations of the dynamic stability and recovery of such shapes. Nonlinear, unstable variations of aerodynamic forces and moments with angle of attack and sideslip were known to exist for these configurations, and extensive conventional force tests, dynamic free-flight model tests, and analytical studies were conducted to define the nature of problems that might be encountered during atmospheric reentry. At Ames, the supersonic and hypersonic free-flight aerodynamic facilities have been used to observe dynamic stability characteristics, extract aerodynamic data from flight tests, provide stabilizing concepts, and develop mathematical models for flight simulation at hypersonic and supersonic speeds.

Meanwhile, at Langley, researchers in the Spin Tunnel were conducting dynamic stability investigations of the Mercury, Gemini, and Apollo capsules in vertically descending subsonic flight.[56] Results of these studies dramatically illustrated potential dynamic stability issues during spacecraft recovery. For example, the Gemini capsule model was unstable; it would at various times oscillate, tumble, or spin about a vertical axis, with its symmetrical axis tilted as much as 90 degrees from the vertical. However, the deployment of a drogue parachute during any spinning or tumbling motions quickly terminated these unstable motions at subsonic speeds. Extensive tests of various drogue-parachute configurations resulted in definitions of acceptable parachute bridle-line lengths and attachment points. Spin tunnel results for the Apollo Command Module configuration were even more dramatic. The Apollo capsule with blunt end forward was dynamically unstable and displayed violent gyrations, including large oscillations, tumbling, and spinning motions. With the apex end forward, the capsule was dynamically stable and would trim at an angle of attack of about 40 degrees and glide in large circles. Once again, the use of a

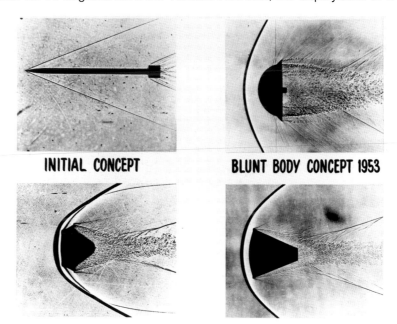

INITIAL CONCEPT   **BLUNT BODY CONCEPT 1953**

Relevant research in dynamic free-flight model research in the Ames Supersonic Free Flight Facility supporting Project Mercury. Similar research efforts have supported many other capsule configurations, including tests of the current Orion capsule shape in the Ames Hypervelocity Free-Flight Facility.

drogue parachute stabilized the capsule, and the researchers also found that retention of the launch escape system, with either a drogue parachute or canard surfaces attached to it, would prevent an unacceptable apex-forward trim condition during launch abort.

After the Apollo program, NASA conducted a considerable effort on unpiloted space probes and planetary exploration. The packaging of instruments designed to collect information on planetary atmospheres involved the use of capsules designed to survive entry into a planetary atmosphere and to descend in a stable condition to the planet's surface. In the Langley Spin Tunnel, several planetary-entry capsule configurations were tested to evaluate their dynamic stability during descent, with a priority in simulating descent in the Martian atmosphere.[57] Simulation of this scenario required consideration of new scaling procedures for dynamic models, because the piloted capsule research just discussed was conducted in the Earth's gravitational field, which was common to both the models and the full-scale articles. In addition, the thinner atmosphere of Mars represented a challenge in simulating a scaled rate of descent. After studies of the simulation requirements, it was concluded that the scaling laws used for dynamic models would not have to be changed, and that the correction in results for the different gravitational fields could be applied when the data were converted from model-scale to full-scale values. Studies also included assessments of the Pioneer Venus probe in the 1970s. These tests provided considerable design information on the dynamic stability of a variety of potential planetary exploration capsule shapes. Additional studies of the stability characteristics of blunt, large-angle capsules were conducted in the late 1990s in the Spin Tunnel.[58] The investigation included four entry-vehicle shapes with variable center-of-gravity locations within the test variables.

Photograph of the free-flight Mercury capsule in vertical descent in the Spin Tunnel with drogue parachute deployed. Tests to determine the dynamic stability characteristics of capsules have continued to this day.

As the new millennium began, NASA's interests in piloted and unpiloted planetary exploration resulted in additional studies of dynamic stability in the Spin Tunnel. Studies have included the Stardust comet probe and the proposed Mars and Moonrise lunar sample return missions. In the Stardust study, the results of the investigation first demonstrated the effectiveness of a drogue parachute for enhanced stability, and the required parachute area, riser-line length, and bridle geometry were defined for subsonic stability, but the geometry was also applied for supersonic conditions to damp oscillations. Sample return vehicles have been investigated for their dynamic behavior, which has included unexpected dynamic instabilities exhibited by statically stable configurations.[59] Modern parameter identification (PID) techniques for extraction of aerodynamic data from flight tests have been factored into the investigations to derive inputs for numerical simulation of descent. The tunnel and its dynamic model testing techniques are supporting NASA's Constellation program for lunar exploration. Included in the dynamic stability testing are the Orion launch abort vehicle, the crew module, and alternate launch abort systems.[60] Once again, the facility is providing guidance for the design of drogue parachutes to stabilize the reentry of the Orion capsule.[61]

Reentry Vehicles and Lifting Bodies

The rapid accelerations in aircraft speed and altitude capabilities that occurred at the end of World War II precipitated the beginning of serious interest in the possibility of piloted vehicles for access to and recovery from space for military and civil missions. Having conquered the transonic and supersonic flight regimes, researchers turned their attention to hypersonic flight and its challenges, which included propulsion systems, structural heat loads, mission operational issues, and handling qualities of the unconventional vehicles under consideration. The NACA and military visionaries initiated early efforts for the X-15 hypersonic research aircraft, in-house design studies for hypersonic vehicles were started at Langley and Ames, and the Air Force began its X-20 Dyna-Soar space plane program. The evolution of long, slender configurations and others with highly swept lifting surfaces were yet another perturbation of new and unusual vehicles with unconventional aerodynamic, stability, and control characteristics requiring free-flight models for assessments of flight dynamics.

In addition to the high-speed studies of the X-15 in the Ames supersonic free-flight facility previously discussed, the X-15 program sponsored low-speed investigations of free-flight models at Langley in the Full-Scale Tunnel, the Spin Tunnel, and an outdoor drop model launched from a helicopter.[62] The Full-Scale Tunnel study concentrated on assessments of the high-angle-of-attack behavior of the early X-15 configuration (known as configuration 1), which included a tall vertical tail and fuselage side fairings that extended to the nose of the airplane. Key X-15 features under study were the use of an all-moving vertical tail for yaw control and differential deflection of the all-moving horizontal tail surfaces for roll control. The flight tests were conducted for angles of attack as high as 30 degrees, where the longitudinal and directional stability were noticeably degraded; however, the effectiveness and control harmony provided by the vertical tail and differential tail were satisfactory. Additional free-flight studies were made in the Full-Scale Tunnel for the final X-15 design (configuration 3), which did not have extended fuselage side fairings and used a symmetrical vertical tail configuration with a jettisonable lower rudder for landing. The results were similar to those previously obtained and provided confidence in the design.

The most significant contribution of the NASA free-flight tests of the X-15 was confirmation of the effectiveness of the differential tail for control. North American had followed pioneering research at Langley on the use of the tail for roll control and had used such a design in its YF-107A aircraft. It opted to use the concept for the X-15 to avoid ailerons that would have complicated wing design for the hypersonic aircraft. Nonetheless, skepticism existed over the potential effectiveness of the application until the free-flight tests at Langley dramatically demonstrated its success.[63]

Along with the X-15 program, NASA activities in the late 1950s included a broad research program on hypersonic glider designs. At Langley, many concepts were conceived by performance-oriented organizations and explored with free-flight models in the Full-Scale Tunnel, including half cones, pyramid-shaped vehicles, lenticular shapes, variable-geometry designs with foldout wings, and flat-bottom configurations with high wing sweep (on the order of 80 degrees).[64] Throughout these studies, results were compared to experiences with the X-15 to establish the feasibility of the unorthodox shapes for operational missions. On the basis of lift-to-drag ratio and wing loadings, the aerodynamic gliding performance of the concepts was acceptable, and the longitudinal characteristics of the designs appeared satisfactory, but the lateral-directional behavior of the highly swept winged designs was unsatisfactory without artificial roll stabilization. Specifically, the models exhibited large unacceptable roll oscillations at moderate and high angles of attack. As the existence of the oscillations was known, it was possible to design lateral control systems that damped the motions and produced satisfactory characteristics. In addition to the in-house exploratory studies, Langley organizations contributed to the development of the short-lived Air Force Dyna-Soar project, including free-flight model tests in the Langley Full-Scale Tunnel.[65]

The most important contributions of free-flight models to reentry gliders occurred while the full-scale X-15 was being flight-tested. These historic efforts were produced during the conception and development of wingless lifting bodies. In the late 1950s, scientists at NASA Ames conducted in-depth studies of the aerodynamic and aerothermal challenges of hypersonic reentry and concluded that blunted half-cone shapes could provide adequate thermal protection for vehicle structures while also producing a significant expansion in operational range and landing options. As interest in the concept intensified after a major conference in 1958, a series of half-cone free-flight models provided proof that such vehicles exhibited satisfactory flight behavior.

Photograph of the Air Force Dyna-Soar configuration in free flight in the Full-Scale Tunnel. The project contributed critical information on lateral-directional stability as part of a broad research program within NASA on the low-speed dynamic stability and control characteristics of hypersonic vehicles.

At Langley, researchers had followed the development of an Ames reentry concept known as the M1 configuration, which resembled the nose of a blunted rocket. Intrigued by the issues concerning stability and control of this radical shape, the staff of the Full-Scale Tunnel conducted free-flight tests of a one-third-scale M1 model in 1959 to develop control surface concepts and to assess the dynamic stability and control of the configuration. After major control deficiencies were encountered in the test program, a modified configuration produced positive results that further fueled the interest in lifting bodies.[66]

The most famous free-flight model activity in support of lifting body development was stimulated by the advocacy and leadership of Dale Reed of the Dryden Flight Research Center. In 1962, Reed became fascinated with the lifting body concept and proposed that a piloted research vehicle be used to validate the potential of lifting bodies.[67] He was particularly interested in the flight characteristics of a second-generation Ames lifting body design known as the M2-F1 concept. After Reed's convincing flights of radio-controlled models of the M2-F1, ranging from kite-like tows to launches from a larger radio-controlled mother ship, demonstrated the satisfactory flight characteristics of the M2-F1, Reed obtained approval for the construction and flight-testing of his vision of a low-cost piloted unpowered glider. The motion-picture films of Reed's free-flight model flight tests had an overwhelming effect on skeptics, and management's support led to a decade of successful lifting body flight research at Dryden.

At Langley, support for the M2-F1 flight program included free-flight tow tests of a model in the Full-Scale Tunnel, and the emergence of Langley's own lifting body design, known as the HL-10, resulted in wind tunnel tests in nearly every facility at Langley.[68] Free-flight testing of a dynamic model of the HL-10 in the Full-Scale Tunnel demonstrated outstanding dynamic

Photograph of the Ames M1 reentry concept in free-flight during evaluations in 1959. The model was powered by compressed air through a nozzle at the rear of the model and used two pairs of control surfaces.

Photograph of Dryden free-flight research models of reentry lifting bodies. Dale Reed, second from left, and his test team pose with the mother ship and models of the M2-F2 and the Hyper III configurations.

stability and control to angles of attack as high as 45 degrees, and rolling oscillations that had been exhibited by the earlier highly swept reentry bodies were completely damped for the HL-10 with three vertical fins.[69] Within its lifting body activities, Langley also responded to an Air Force request to evaluate the stability of the Precision Recovery Including Maneuvering Entry (PRIME) SV-5 unpiloted reentry vehicle on tow during retrieval using a C-130 aircraft.[70]

In the early 1970s, a new class of lifting body dubbed "racehorses" by Dale Reed emerged.[71] Characterized by high fineness ratios, long pointed noses, and flat bottoms, these configurations were much more efficient at hypersonic speeds than had been the earlier "flying bathtubs." Reed and his team evaluated one Langley-developed configuration, known as the Hyper III, using free-flight models and the mother ship test technique. Although the Hyper III was efficient at high speeds, it exhibited a low lift-to-drag ratio at low speeds, requiring some form of variable geometry such as a pivot wing, flexible wing, or gliding parachute.

Reed successfully advocated for a low-cost, 32-foot-long helicopter-launched demonstration vehicle of the Hyper III with a pop-out wing, which made its first flight in 1969. Flown from a ground-based cockpit, the Hyper III flight was launched from a helicopter at an altitude of 10,000 feet. After being flown in research maneuvers by a research pilot using instruments, the vehicle was handed off to a safety pilot, who landed it. Unfortunately, funding for a low-cost piloted project similar to the earlier M2-F1 activity was not forthcoming for the Hyper III.

Ultimately, the NASA Space Shuttle configuration was selected as a winged vehicle in lieu of derivative lifting body options. Nonetheless, the successful lifting body concept has been revisited many times by NASA and international space programs. The NASA lifting body follow-on programs have included the HL-20 lifting body space ferry, the X-30 National Aero-Space Plane, the X-33 Shuttle replacement, and the X-38 Crew Return Vehicle, which used a parafoil gliding parachute for landings.

During 1992, a Dryden team including Reed demonstrated the viability of an autonomous landing concept proposed for the X-38, using Global Positioning System (GPS) for navigation and maneuvering to a landing site. In the Spacecraft Autoland project, a 4-foot-long flattened biconical generic airframe called Spacewedge was used with a ram-air parafoil and controlled by a small onboard computer for autonomous flight.[72] Using the vehicle shape of the Air Force–NASA X-24 lifting body, the X-38 team subsequently conducted radio-controlled drop tests of a 4-foot-long model equipped with a ram-air parafoil launched from a general aviation airplane in 1995. In 2000, a 0.80-scale X-38 vehicle was released from a B-52 at an altitude of 39,000 feet and completed a successful landing, but the X-38 program was canceled in 2002.

The full-scale version of the Hyper III was 32 feet long, launched from a helicopter at 10,000 feet, and controlled by a NASA research pilot from a ground-based cockpit.

Supersonic Transports

The commitment of the United States in 1963 to develop a civil Supersonic Transport had a major impact on NASA's aeronautical centers. Nearly every discipline in aerospace technology was challenged to supply design data and assessments of feasible aircraft configurations. Within aerodynamics, substantial advances in the state of the art for supersonic and transonic performance were made; however, the highly swept wing planforms used by most of the candidate designs were expected to exhibit poor low-speed characteristics at take-off and landing conditions. Dynamic free-flight models were considered a necessity to evaluate potential low-speed problems, and the Langley Full-Scale Tunnel was a major participant in the program. Initially, the free-flight model test programs were directed at evaluating the flying characteristics of the competing designs. The original Boeing SST variable-sweep design and Lockheed SST double-delta design were flown extensively in the Full-Scale Tunnel and used in several entries to measure static and dynamic stability and control aerodynamic data.[73] The resulting aerodynamic data were used for analysis of the model flight-test results and for inputs to piloted simulators. The Boeing design showed satisfactory characteristics, except for a marked deterioration in lateral control effectiveness and roll damping when the wing was at maximum sweep (72 degrees). The Lockheed design was also satisfactory, except for extreme angles of attack (28 degrees), where it exhibited a divergence in yaw.

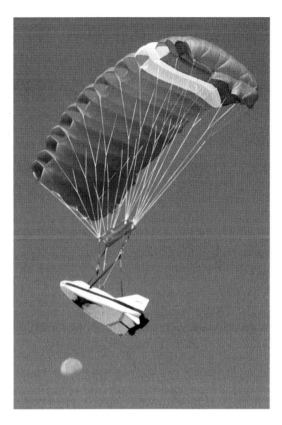

A 0.80-scale model of the X-38 Crew Return Vehicle in controlled glide using a parafoil for landing.

NASA had conducted in-house design studies of Supersonic Transport configurations, which industry evaluated and then selected the most feasible vehicles for technology assessments. One of these configurations, known as the SCAT-15, used auxiliary variable-sweep wing panels for improved low-speed aerodynamic behavior. However, industry did not view favorably the weight penalties associated with the variable-sweep feature, and the NASA team removed the variable-sweep feature and redesigned the configuration, retaining its outstanding supersonic performance in a derivative configuration known as the SCAT-15F design.[74] Unfortunately, the SCAT-15F exhibited longitudinal instability (pitch-up) at moderate angles of attack

Free-flight model of the Boeing variable-sweep model 733 Supersonic Transport undergoes assessments of dynamic stability and control in the Full-Scale Tunnel. After these tests, Boeing changed its configuration for the SST to a fixed wing design.

Free-flight model of the NASA Langley SCAT 15F Supersonic Transport in the Full-Scale Tunnel. The SCAT-15 design still represents the most efficient Supersonic Transport configuration.

and the potential for a dangerous unrecoverable deep-stall trim condition. Many wind tunnel entries were conducted at several Langley facilities in attempts to cure the pitch-up problem, and the free-flight model was used to define the effects of airframe geometric modifications. No acceptable combination of airframe changes was found for the problem, and the lack of sophisticated fly-by-wire control systems routinely used in today's inherently unstable military and civil aircraft was a major blow to the configuration. Despite being analyzed by Boeing under requirements by the FAA, the SCAT-15F was not in contention for the final U.S. SST of the 1970s. The airframe design of the SCAT-15F still represents the most efficient Supersonic Transport configuration.

Vehicle/Store Separation

One of the more complex and challenging areas in aerospace technology is the prediction of paths of aircraft components after the release of items such as external stores, canopies, crew modules, or vehicles dropped from mother ships. Aerodynamic interference phenomena between vehicles can cause major safety-of-flight issues, resulting in catastrophic impact of the components with the airplane, and unexpected pressures and shock waves can dramatically change the expected trajectory of stores. Conventional wind tunnel tests used to obtain aerodynamic inputs for calculations of separation trajectories must cover a range of test parameters, and the requirement for dynamic aerodynamic information further complicates the task. Measurement of aerodynamic pressures, forces, and moments on vehicles in proximity in wind tun-

nels is a challenging technical procedure. The use of dynamically scaled free-flight models can quickly provide a qualitative indication of separation dynamics, thereby providing guidance for wind tunnel test planning and early identification of potentially critical flight conditions.

Separation testing for military aircraft components using dynamic models at Langley evolved into a specialty at the Langley 300 mph 7- by 10-Foot Tunnel, where subsonic separation studies included assessments of the trajectories taken by released cockpit capsules, stores, and canopies. In addition, bomb releases were simulated for several bomb-bay configurations, and the trajectories of model rockets fired from the wingtips of models were also evaluated. As requests for specific separation studies mounted, the staff rapidly accumulated expertise in testing techniques for separation clearance.[75]

One of the more important separation studies conducted in the Langley tunnel was an assessment of the launch dynamics of the X-15/B-52 combination for launches of the X-15. Before the X-15, launches of research aircraft from carrier aircraft had only been made from the fuselage centerline location of the mother ship, and in view of the asymmetrical location of the X-15 under the right wing of the B-52, concern arose as to the aerodynamic loads encountered during separation and the safety of the launching procedure. Separation studies were therefore conducted in the Langley 300 mph 7- by 10-Foot Tunnel and the Langley High-Speed 7- by 10-Foot Tunnel.[76] Detailed measurements of the aerodynamic loads on the X-15 in proximity to the B-52 under its right wing were made during conventional force tests in the High-Speed

Time sequence of a canopy ejection test in the Langley 300 mph 7- by 10-Foot Wind Tunnel. The staff of the facility conducted extensive separation studies of aircraft stores and components using a net to catch components released from models under test.

Free-flight drop model of the X-15 research aircraft undergoes separation testing beneath a B-52 model in the Langley 300 mph Low-Speed 7- by 10-Foot Tunnel. The model was mounted under the left wing of the bomber for these tests for visibility.

Tunnel, while the trajectory of a dynamically scaled X-15 model was observed during a separate investigation in the Low-Speed Tunnel. The test set up for the low-speed drop tests used a dynamically scaled X-15 model under the left wing of the B-52 model to accommodate viewing stations in the tunnel. The model used conventional Froude-number scaling procedures so that the model and full-scale aircraft translational accelerations were equal; therefore, the effects of Mach number could not be determined. Initial trim settings for the X-15 were determined to avoid contact with the B-52, and the drop tests showed that the resulting trajectory motions provided adequate clearance for all conditions investigated.

During successful subsonic separation events, a bomb or external store is released, and gravity typically pulls it away safely. At supersonic speeds, however, aerodynamic forces are appreciably higher relative to the store weight, shock waves may cause unexpected pressures that severely influence the store trajectory or bomb guidance system, and aerodynamic interference effects may cause catastrophic collisions after launch. Under some conditions, bombs released from within a fuselage bomb bay at supersonic speeds have encountered adverse flow fields, to the extent that the bombs have reentered the bomb bay. In the early 1950s, the NACA advisory committees strongly recommended that focused efforts be initiated by the Agency in store separation, especially for supersonic flight conditions. Langley responded with an investigation of supersonic bomb releases at a Mach number of 1.62 in the 9-Inch Supersonic Tunnel.[77] The dynamic model test examined the effect of various store locations on a swept wing and several bomb-bay configurations. Results of the study were among the first to illustrate the adverse interference effects of supersonic releases.

Researchers within Langley's Pilotless Aircraft Research Division (PARD) used their Preflight Jet facility at Wallops to conduct research on supersonic separation characteristics for several high-priority military programs.[78] The preflight facility was designed to check out ramjet engines before rocket launches, consisting of a blow-down–type tunnel powered by compressed air exhausted through a supersonic nozzle. The main jet was square in cross section and was constructed in two sizes: 12 inches and 27 inches. Test Mach number capability was from 1.4 to 2.25. With an open throat and no danger to a downstream facility drive system, the facility proved ideal for dynamic studies of bombs or stores after supersonic releases. PARD devised appropriate scaling laws for the simulation of Mach number and conducted extensive separation tests.[79]

One of the more crucial tests conducted in the Wallops Preflight Jet facility was support for the development of the Republic F-105 fighter bomber, which was specifically designed with forcible ejection of bombs from within the bomb bay to avoid the issues associated with external releases at supersonic speeds. For the test program, a half-fuselage model (with bomb bay) was mounted to the top of the nozzle, and the ejection sequence included extension of folding fins on the store after release. A piston and rod assembly forcefully ejected the store from the open bomb bay, and high-speed photography documented the motion and trajectory of the store. The F-105 program expanded to include numerous specific and generic bomb and store shapes requiring almost 2 years of tests in the facility. The ejection tests in the Wallops facility were supplemented by force and moment measurements in the Langley 4-Foot Supersonic Tunnel, where the aerodynamic loads acting on the store were measured for various positions of the store relative to the aircraft. Numerous generic and specific aircraft separation studies in the Preflight Jet facility from 1954 to 1959 included F-105 pilot escape, F-104 wing drop-tank separations, F-106 store releases from an internal bomb bay, and B-58 pod drops.

The national challenge in advancing the state of the art in store separation dynamics was ultimately taken over by the research and development establishments within the Air Force and Navy, and NASA's role was minimized after the 1960s.

Advanced Civil Aircraft

Most of the free-flight model research conducted by NASA to evaluate dynamic stability and control within the flight envelope has focused on military configurations and a few radical civil aviation designs. This situation resulted from advances in the state of the art for design methods for conventional subsonic configurations over the years and many experiences correlating results of model and airplane tests. As a result, transport design teams have collected massive data and experience bases for transports, which serve as the corporate knowledge base for derivative aircraft. For example, companies now have considerable experience with the accuracy of their conventional static wind tunnel model tests for the prediction of full-scale aircraft characteristics, including the effects of Reynolds number. Consequently, testing techniques such as free-flight tests do not have high technical priority for such organizations. Conventional civil transport configurations have therefore not been test subjects in NASA free-flight studies, with the exception of special situations, such as poststall motions or wake vortex encounters, to be discussed in later sections.

The radical Blended Wing-Body (BWB) flying wing configuration has been a notable exception to the foregoing trend. Initiated with NASA sponsorship at McDonnell-Douglas (now Boeing) in 1993, the subsonic BWB concept carries passengers or payload within its wing structure to minimize drag and maximize aerodynamic efficiency.[80] Over the past 16 years, wind tunnel research and computational studies of various BWB configurations have been conducted by NASA–Boeing teams to assess cruise conditions at high

subsonic speeds, take-off and landing characteristics, spinning and tumbling tendencies, emergency spin/tumble recovery parachute systems, and dynamic stability and control. In 2000, NASA and Boeing initiated a flight project to design, fabricate, and fly a 35-foot-span remotely piloted model of a BWB variant known as the X-48A, with flight-testing to take place at Dryden. After a year of startup activity, the program was canceled by NASA, which redirected its research thrusts toward a 12-foot-span wind tunnel free-flight model designed for the Full-Scale Tunnel.

By 2005, the BWB team had conducted static and dynamic force tests of models in the 12-Foot Low-Speed Tunnel and the 14-by 22-Foot Tunnel to define aerodynamic data used to develop control laws and

The Boeing X-48B Blended Wing-Body configuration in flight at NASA Dryden. The configuration has undergone almost 15 years of research, including free-flight testing at Langley and Dryden. The stinger visible at the rear of the model is used for deployment of a parachute for emergency spin and tumble recovery.

control limits, as well as trade studies of various control effectors available on the trailing edge of the wing. Free-flight testing then occurred in the Full-Scale Tunnel with the 12-foot-span model.[81] Results of the flight test indicated satisfactory flight behavior, including assessments of engine-out asymmetric thrust conditions.

In 2002, Boeing contracted with Cranfield Aerospace, Ltd., for the design and production of a pair of 21-foot-span remotely piloted models of BWB vehicles known as the X-48B configuration. After conventional wind tunnel tests of the first X-48B vehicle in the Langley Full-Scale Tunnel in 2006, the second X-48B underwent its first flight in July 2007 at the Dryden Flight Research Center. The BWB flight-test team is a cooperative venture between NASA, Boeing Phantom Works, and the Air Force Research Laboratory. The first 11 flight tests of the 8.5-percent-scale vehicle in 2007 focused on low-speed dynamic stability and control with wing leading-edge slats deployed. In a second series of flights, which began in April 2008, the slats were retracted, and higher-speed studies were conducted. Powered by three model aircraft turbojet engines, the 500-pound X-48B is expected to have a top speed of about 140 mph. A sequence of flight phases is scheduled for the X-48B, with various objectives within each study directed at the technology issues facing the implementation of the innovative concept. The application focus for the BWB configuration has changed from the original vision of a civil transport to a long-range, high-capacity military transport or tanker.

NASA's interest in flight dynamics studies for smaller personal-owner aircraft has been directed at unconventional configurations that offer advantages in performance, stability, control, or safety. In the 1980s, Langley conducted several cooperative studies with famous designer Elbert "Burt" Rutan to explore the benefits offered by canard-type general-aviation aircraft. One particular advantage, which has been demonstrated by Rutan's designs, is the inherent stall-proof nature of properly designed canard aircraft.[82] The scope of the studies included full-scale conventional wind tunnel tests of Rutan's Varieze design, NASA pilot evaluations of the flight characteristics of the full-scale Varieze and twin-engine Defiant airplanes, and free-flight tests of a dynamic model of the Varieze in the Full-Scale Tunnel.[83] The database gathered in the studies significantly increased the design information and understanding of the stall- and spin-resistant qualities of these successful canard configurations.

Wake Vortex Hazard

Since the introduction of large, wide-body civil transports in the mid-1960s, considerable research and development has been directed at the potential hazards involved in encounters by light aircraft with the trailing vortex wake shed by generator aircraft in the terminal area. The intensity of the vortical flows, which is directly related to the weight of the generator aircraft, can result in catastrophic uncontrolled excursions at low altitudes for smaller aircraft attempting to penetrate the wake. Regulatory separation distances have been invoked to ensure that such encounters are minimized or eliminated during routine airport operations; however, they impose a primary constraint on airport throughput and capacity. NASA has conducted research on the wake vortex hazard at Langley and Dryden since the early 1970s in a broad program that includes aircraft flight tests to measure the strength and duration of vortex wakes, wind tunnel and flight tests to conceive and evaluate the effects of airframe modifications on the problem, computational aerodynamic studies to provide guidance for modifications and understanding of vortex properties, and consultation with regulatory agencies.

As part of this ongoing effort, Langley conducted an exploratory free-flight model study in the Full-Scale Tunnel in 1995 to determine whether the free-flight test technique was useful in wake vortex encounter research.[84] Objectives of the study were to determine if a free-flight model could be safely and accurately flown into wake vortexes generated by a stationary wing, to estimate the rolling moments imposed on the following model, and to determine if the relative severity of the encounter event could be categorized depending on the initial encounter conditions. The arrangement for the free-flight tests consisted of a fixed vortex-generating wing in a forward position within the test section and a generic business-class free-flight transport model. Smoke was used to define the trailing vortex system behind the generator wing in the pilots of the following model attempted to fly into the vortex from various trajectories. The trailing model was instrumented with a three-axis rate gyro, a three-axis accelerometer, control position potentiometers, and wingtip booms to measure local angle of attack and sideslip. Results of the study indicated that the wind tunnel free-flight test technique could be used to fly a model in the vicinity of trailing vortexes, that approximate values of the induced rolling moment on the following model could be determined, and that the different responses of the following model to encounter trajectories and vortex strengths could be assessed. Based on the success of these exploratory tests, follow-on tests were conducted in 1995 (unpublished) to assess the effect of vortex encounters on a twin-engine jet transport configuration.

Free-flight model of the Rutan Varieze canard airplane undergoes assessments of dynamic stability and control in the Full-Scale Tunnel. NASA researchers investigated the configuration for conditions far beyond its realistic flight envelope to gather data for canard-type aircraft.

Towed Vehicles

Yet another application of free-flight models has been for assessments of the dynamic stability of towed vehicles. Depending on the tow line and attachment point configurations, some vehicles exhibit violent instabilities during towed flight, whereas they are

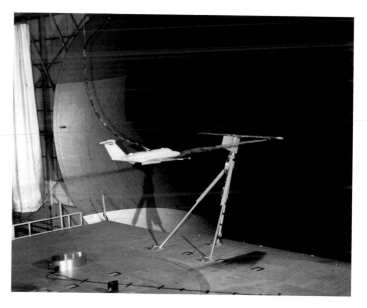

Free-flight model of a generic business-class airplane flies into the tip vortex generated by an isolated wing as part of a study of the application of free-flight testing to studies of the safety of flight issue in the Full-Scale Tunnel.

dynamically stable and well-behaved in free flight without a tow line. The NACA and NASA received many requests from the military for investigations of towed configurations using free-flight model techniques in the 12-Foot Free-Flight Tunnel and the Full-Scale Tunnel. The scope of towed-vehicle studies included gliders, air-to-air targets, minesweepers, parawing cargo delivery systems, parasitic fighters, objects on parachutes designed for midair capture and retrieval, and lifting bodies.[85]

Towing studies started in the Free-Flight Tunnel in 1943 with experimental tests of a glider towed by twin towlines accompanied by a theoretical study of the stability of the configuration.[86] Later during the war, the Army/Air Forces requested support testing of the Cornelius XFG-1 towed fuel glider and potential tow arrangements for the Lockheed P-80A aircraft.[87] More glider requests came from the Navy in 1948.[88] Additional testing was provided in response to an Air Force request for dynamic stability characteristics of the Republic F-84E under tow and a similar Navy request for the Chance Vought F7U Cutlass in towed flight.[89]

The author observes low-speed tow stability tests of a free-flight model of the Ames M2-F1 lifting body in the Full-Scale Tunnel in 1962.

Concept Demonstrators

The efforts of the NACA and NASA in developing and applying dynamically scaled free-flight model testing techniques have matured impressively. Although the scaling relationships have remained constant since the inception of free-flight testing, the facilities and test attributes have become dramatically more sophisticated. The size and construction of models have changed from unpowered balsa models weighing a few ounces with wingspans of less than 2 feet to large powered composite models weighing over 1,000 pounds. Control systems have changed from simple solenoid bang-bang controls operated by a pilot with visual cues provided by model motions to hydraulic systems with digital flight controls and full feedbacks from an array of sensors and adaptive control systems. The level of sophistication integrated into the model testing techniques has given rise to a class of free-flight models that are considered to be integrated concept demonstrators rather than specific technology tools. Thus, the line between free-flight models and more complex remotely piloted vehicles has become blurred, with a noticeable degree of refinement in the concept demonstrators.

Research activities at the NASA Dryden Flight Research Center illustrate how far free-flight testing has come. Since the 1970s, Dryden has conducted a broad program of demonstrator applications with emphasis on integrations of advanced technology. In addition to the previously discussed X-48B program, two other Dryden projects, the Highly Maneuverable Aircraft Technology (HiMAT) program and the X-36 program, are related to the main theme of this document on research for dynamic stability and control of advanced vehicles.

After the success of the exploratory remotely piloted vehicle project using the three-eighths-scale unpowered F-15 Spin Research Vehicle, Dryden accelerated its efforts in the development of remotely piloted research vehicle (RPRV) technologies in the cooperative HiMAT program with the Air Force Flight Dynamics Laboratory in 1979.[90] The objective was to demonstrate integrated advanced fighter technologies such as advanced composites and aeroelastic tailoring, close-coupled canards, and winglets. Launched from a B-52 mother ship at altitudes of about 45,000 feet, the HiMAT remotely piloted research vehicle was about half the size of an F-16 fighter and was powered by a J85 jet engine. Designed by the team of

The HiMAT remotely piloted research vehicle flies over Edwards Air Force Base. In its role as an advanced technology demonstrator, the vehicle incorporated a highly sophisticated flight control system, a canard, winglets, aeroelastic tailoring, and relaxed static stability.

Ames, Dryden, and Rockwell International, the configuration demonstrated twice the turn rate capability of the F-16 at transonic speeds. An onboard digital computer provided fly-by-wire control capabilities, including propulsive control. The vehicle was designed with relaxed static stability (from 10- to 30-percent unstable) and direct force control, and active control was a major area of emphasis in the research flights. Two HiMAT vehicles conducted 26 flight tests during the program.

The HiMAT program produced extensive information regarding the flight dynamics characteristics of a radically new configuration for highly maneuverable aircraft. In addition to results on aeroelastic tailoring, automated flight maneuvers for extraction of aerodynamic data, and remote-piloting technologies for supersonic aircraft, the study inspired aircraft design features. For example, designers of the Grumman X-29 forward-swept wing research aircraft employed much of the technology from the HiMAT experiments, including a close-coupled canard, relaxed static stability, and lightweight composite materials.[91]

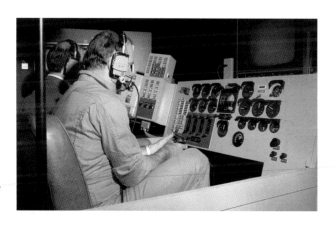

NASA research pilot Bill Dana flies the HiMAT vehicle from a fixed-base ground cockpit.

In 1997, another milestone was achieved at Dryden in remotely piloted research vehicle technology when a NASA–Boeing X-36 vehicle demonstrated the feasibility of using advanced technologies to ensure satisfactory flying qualities for radical tailless fighter designs. The X-36 was designed as a joint effort between the NASA Ames Research Center and Boeing Phantom Works (previously McDonnell-Douglas) as a 0.28-scale powered free-flight model of an advanced fighter without vertical or horizontal tails to enhance survivability. Powered by a F112 turbofan engine and weighing about 1,200 pounds, the 18-foot-long configuration used a canard, split aileron surfaces, wing leading- and trailing-edge flaps, and a thrust-vectoring nozzle for control. A single-channel digital fly-by-wire system provided artificial stability for the configuration,

The X-36 tailless advanced fighter flight demonstrator in flight at NASA Dryden. The radical configuration conducted a successful flight investigation in one of the more remarkable remotely piloted vehicle programs at Dryden.

which was inherently unstable about the pitch and yaw axes. For the first time, a microphone was carried onboard a powered free-flight model in the cockpit area of the X-36. The aural cues provided alerts to the pilot for engine stalls or other propulsion-related problems. The ground-based pilot used a fighter-type cockpit outfitted with a standard head-up display and a moving-map representation of the location of the X-36 in a successful series of 31 subsonic research flights, during which the vehicle reached an altitude of about 20,000 feet and a maximum angle of attack of 40 degrees. The pilot's crew station was in a ground control station trailer, where distractions were minimized.[92]

Summary

This overview of over 80 years of applications of free-flight dynamically scaled models for assessments of the dynamic stability and control characteristics of full-scale vehicles has discussed the remarkable advances that the NACA and NASA have made in conducting and correlating results of the studies. The impact of the model results for advanced military and civil aircraft, spacecraft, and utility vehicles has influenced the development of new aircraft, provided an early identification of problems and potential solutions, and reduced the risk of subsequent full-scale flight tests.

CHAPTER 5:

FLIGHT AT HIGH ANGLES OF ATTACK

The first efforts in aircraft stability and control concentrated on providing satisfactory characteristics within the flight envelope, especially for cruise conditions. Rapid progress was made in the development and validation of theoretical prediction methods, test procedures and the interpretation of conventional wind tunnel test data, flight-test procedures for full-scale aircraft, and the development and use of dynamically scaled free-flight models. However, from the earliest days of heavier-than-air flying machines, it was recognized that critical conditions were encountered during flight at high angles of attack near wing stall, where flow separation on the wing might create unacceptable—even catastrophic—stability and control problems. Such problems often occurred at low altitudes during take-off and landing, resulting in loss of control with insufficient altitude for recovery. Analysis of the critical factors that influenced flight behavior at high angles of attack was difficult because of the nature of the complex separated-flow aerodynamic phenomena encountered near stall and higher angles of attack. As a consequence of the poor state of aerodynamic understanding, design methods were essentially nonexistent. An alarming increase in fatal stall/spin accidents in the 1930s resulted in a high priority being placed on obtaining reliable predictions early in the design process. As a result of the relative immaturity of theoretical methods, the application of dynamically scaled free-flight models for research on high-angle-of-attack flight behavior became recommended for general research and evaluations of specific aircraft. Applications of dynamic models by NASA in this area have continued to the present day.

Early NACA free-flight studies in the Langley 12-Foot Free-Flight Tunnel routinely included assessments of the dynamic stability and controllability of models for a range of angles of attack, including values associated with maximum lift and stall. In the 1930s and 1940s, aircraft configurations rarely used swept-back wings, and the primary problems observed in the tunnel tests for high-angle-of-attack conditions typically included degraded roll control effectiveness caused by adverse yawing moments created by aileron deflections and abrupt, uncontrollable roll-off at stall caused by sudden asymmetric stalling of the wing. On occasion, configurations exhibited lightly damped Dutch roll lateral-directional oscillations with unacceptable rolling and yawing excursions at high angles of attack.

When early NACA research on the stability and control characteristics of swept wing configurations encountered problems caused by tip stall at high angles of attack, several radical concepts were evaluated to minimize the problem. This free-flight model was used for flight investigations of a "W" wing configuration with unswept outer wing panels in an attempt to alleviate tip stall.

As previously discussed, results obtained with the early free-flight models were correlated with full-scale aircraft flight results to establish the validity of the testing technique. In most cases, the relatively low Reynolds number of the model tests resulted in a lower magnitude of aerodynamic moments produced by control surfaces and a lower angle of attack for stall. Nonetheless, the model tests were viewed as a reliable qualitative indicator of gross problems, and free-flight test results provided a valuable awareness of potential problems. The tests also permitted a rapid assessment of candidate solutions before full-scale flight tests. In general, the free-flight models exhibited characteristics similar to the full-scale aircraft during high-angle-of-attack flight conditions and stalls; however, the model behavior usually occurred at a slightly lower angle of attack.[1]

During the late 1940s and 1950s, the introduction of radical configurations with swept and delta wings resulted in more issues for flight at high angles of attack. The use of sweptback wings resulted in spanwise flow over the wing at moderate and high angles of attack, causing separated flow on the outer wing and dramatic losses of aileron effectiveness and longitudinal instability ("pitch-up") within the intended operational envelope. Wing sweep also produced excessive levels of effective dihedral for many configurations and resulted in aggravation of Dutch roll tendencies. In response to these challenges, the NACA introduced generic research projects to provide design guidelines for acceptable levels of stability and control for swept wing aircraft, and to conceive and evaluate the effectiveness of geometric modifications (such as wing fences) to minimize or eliminate the adverse flow properties. After intense free-flight model research had been accomplished with generic swept wing configurations, requests began to be received from the Air Force and Navy for evaluations of new fighters such as the McDonnell XF3H-1 Demon (the first swept wing fighter built by McDonnell).[2]

Along with emphasizing effects of sweptback wings on dynamic stability, the NACA researchers in the Free-Flight Tunnel started major research investigations on delta wings. Assessments of the high-angle-of-attack dynamic stability and control characteristics of delta wing configurations in 1948 indicated fairly good characteristics for models having delta wings with sweep-back angles from 50 to 60 degrees, although the relatively low lift-to-drag ratios for such low aspect ratios (2 or 3) resulted in steep power-off glide angles. When the wing leading-edge sweep of the delta wing models was increased to 70 or 80 degrees (aspect ratio of 1 or less), the stability and control characteristics became unsatisfactory because of large, constant-amplitude rolling oscillations at high angles of attack, ineffective roll control, and an abrupt onset of pitch-up instability as angle of attack was increased.[3] These pioneering studies of the dynamic stability and control of swept wing and delta wing free-flight models uncovered physical phenomena and aircraft characteristics that were impossible to predict with conventional static force test techniques. Similar dynamic stability and control characteristics soon became major issues for emerging highly swept- and delta-type aircraft configurations for high-angle-of-attack conditions. After the NACA's basic research was disseminated, the military services requested free-flight tests at Langley for several new delta wing aircraft, including the Douglas XF4D-1 Skyray and the Convair F-102 Delta Dagger.[4]

The investigations conducted in the Free-Flight Tunnel on dynamic stability and control of the early fighters of the 1950s provided confidence that the behavior of swept-and delta-configurations might be satisfactory for operational aircraft and pointed the way to solutions for unexpected problems. In addition, two important general results were identified by the research. One lesson was that aircraft configurations with swept wings having relatively sharp leading edges did not exhibit the large Reynolds number effects on stability and control at high angles of attack (below maximum lift) as had been experienced with the older aircraft designs. Based on correlations with full-scale results and data accumulated over the last 50 years, it appears that free-flight models have exhibited amazingly accurate predictions of high-angle-of-attack behavior for this type of aircraft. An example of the accuracy of low Reynolds number free-flight tests in predicting full-scale aerodynamic stability and control behavior was NASA tests of the F-4 Phantom II fighter.[5] In that research effort, conventional static wind tunnel data were measured in the NASA Ames 12-Foot Pressure Tunnel at high values of Reynolds number and correlated with model force tests in the Free-Flight Tunnel at Langley. For that particular configuration at high-angle-of-attack, subsonic conditions, the free-flight F-4 model tested in the Full-Scale tunnel accurately predicted the angle of attack for maximum lift, lateral and directional stability, and aerodynamic control characteristics.

The second major result was the experience gained in understanding and predicting instabilities at high angles of attack. In particular, coordinated efforts by Langley researchers in several organizations to compare conventional wind tunnel force and moment measurements with the dynamic free-flight test observations led to the development and validation of predictive criteria for two of the more significant instabilities encountered at high angles of attack—pitch-up and directional divergence ("nose slice"). Directional divergence is a condition at high angles of attack in which massive flow separation on the wing reduces the ability of the vertical tail to maintain "weather-vane" stability for the aircraft. Because the separated flow field results in the impingement of low-energy airflow at the tail location, the magnitude of the stabilizing contribution of the tail is reduced. In addition, for certain configurations, the airflow around the separated flow area can result in an adverse flow direction at the rear of the aircraft, such that the contribution of the vertical tail to stability becomes destabilizing. That is, when the aircraft is sideslipped to the left, the airflow at the vertical tail location is from the left, creating a yawing moment to increase the sideslip angle. When nose slice occurs, it can be violent and abrupt, resulting in loss of control, disorientation of the pilot, and spin entry. The staff of the Free-Flight Tunnel and other Langley research organizations cooperatively developed a predictive theoretical method that is still used today to predict the potential presence of directional divergence instability based on conventional static wind tunnel data.[6]

The previously described Free-Flight Tunnel experiences of 1948 with large-amplitude limit-cycle roll oscillations at high angles of attack for highly swept delta wings did not warrant great interest at the time because such configurations were not under widespread consideration. However, as will be discussed in later sections, the susceptibility of highly swept designs to the roll instability phenomenon became important with the advent of hypersonic configurations and high-performance fighter aircraft with chined forebodies.

In 1959, free-flight model testing moved to the Langley Full-Scale Tunnel, with continued emphasis on high-angle-of-attack studies for advanced aerospace vehicles. One of the first high-priority requests came from the Air Force for studies of the dynamic stability and control characteristics of the delta wing Convair B-58 Hustler bomber early in its developmental cycle.[7] Results of the free-flight test in the Full-Scale Tunnel indicated that the B-58 would have generally satisfactory low-speed longitudinal characteristics; however, at high angles of attack near maximum lift (about 30 degrees), the model exhibited a severe directional divergence, implying that the full-scale aircraft might be susceptible to loss of control or spin entry when flown near that condition. Later in its

Free-flight model of the Convair XB-58 bomber is prepared for dynamic stability and control evaluations in the Full-Scale Tunnel.

operational history, the B-58 was involved in several inadvertent spin incidents, some of which were fatal. Langley later conducted catapult-launched model studies of the spin entry and spinning characteristics of the B-58, and its susceptibility to directional divergence as predicted by the free-flight model was confirmed.[8]

In the early 1960s, the focus of free-flight testing in the Full-Scale Tunnel centered on dynamic stability studies of V/STOL aircraft and reentry lifting bodies. High-angle-of-attack behavior was not a critical issue for horizontal-attitude V/STOL configurations, but vertical-attitude tail-sitter designs necessarily transitioned through large ranges of angle of attack during conversion maneuvers, from hovering to conventional forward flight. Lifting reentry configurations also fly at high angles of attack during typical missions on return from space, and the design of such vehicles requires accurate predictions of dynamic stability and control characteristics across the angle-of-attack range from reentry at hypersonic speeds to landings at subsonic speeds. As previously discussed, subsonic free-flight tests were conducted in the Full-Scale Tunnel for the X-15 with emphasis on the stability and control of the vehicle at high angles of attack. The most significant result of the test program was the verification that differential deflections of the horizontal tail could be used for roll control.

Other families of candidate reentry shapes were flown in the Full-Scale Tunnel, including the proposed Air Force X-20 Dyna-Soar hypersonic glider, lifting bodies, and generic NASA-conceived hypersonic con-

figurations. Two common characteristics of all these test subjects were a highly swept lifting surface and a "fuselage heavy" inertial distribution with most of the mass along the fuselage. As a result of these physical properties, reentry vehicles of the glide-landing type were typically sensitive to aerodynamic parameters acting about the roll axis, and most were found to be susceptible to lightly damped or unstable roll oscillations at moderate and high angles of attack.[9] The early detection of lateral instabilities provided a warning that artificial roll damping would be required for satisfactory behavior, and the flight models demonstrated that the motions could be stabilized with artificial damping in roll provided by rate gyroscopes. For some hypersonic boost-glide models, adverse yaw from lateral control surfaces coupled with the roll instabilities to constrain the usable range of angle of attack. Conventional and dynamic force tests of the models provided aerodynamic data used as inputs for simulator and control system studies in other NASA research organizations and industry.

Free-flight tests of highly swept hypersonic boost-glide configurations demonstrated that the designs would exhibit lightly damped, large-amplitude rolling motions at high angles of attack. Artificial roll-rate stabilization eliminated the oscillations and provided good flight behavior. This photograph from 1957, made before the major move of the 12-Foot Free-Flight Tunnel staff to the Full-Scale Tunnel, shows the rudimentary implementation of the flight-test technique with the crew on an open balcony.

The phenomenon of lightly damped roll oscillations at high angles of attack for highly swept configurations had first been identified by the previously discussed 1948 tests of delta wing models with sweep angles of 80 degrees or more in the Free-Flight Tunnel. The experiences with lifting reentry shapes led to more detailed aerodynamic and theoretical analyses and a better understanding of the factors causing the instability. Static and dynamic wind tunnel force tests revealed that a complex, time-dependent vortical-flow phenomenon was created on the upper surfaces of the wings, resulting in high levels of effective dihedral and unstable dynamic roll damping. Over the years, this flow situation has been observed in free-flight tests of many highly swept models and has precipitated additional research to mathematically model the "wing rock" characteristics exhibited by the models.[10]

Fighters

In the late 1950s and early 1960s, NACA and NASA free-flight model testing of fighter configurations had been limited by a strong military perspective that high-angle-of-attack capability was not a primary design requirement. Instead, military planners envisioned future conflicts to be long-range missile engagements in which the combatants did not even make visual contact. Stringent maneuvers and flight at high angles of attack would not be required, and exposure to any potential loss of control would be avoided. At the same time, however, the services required classic intentional spin and recovery demonstrations to ensure satisfactory spin recovery. These military philosophies had a large impact on free-flight testing of fighters at Langley, resulting in separate, largely uncoordinated testing in the Free-Flight Tunnel and the Spin Tunnel. This situation would change in the mid-1960s.

As U.S. military air forces entered the era of the Vietnam conflict, they were faced with a major deficiency in first-line fighter and attack aircraft. In Vietnam, close-in traditional dogfights frequently occurred, requiring extreme maneuverability and good handling characteristics at high angles of attack. Unfortunately, many of the Nation's aircraft, such as the F-4 Phantom II, F-100 Super Sabre, and A-7 Corsair II, exhibited severe deficiencies in lateral-directional stability and control at high angles of attack, resulting in loss of control, unintentional spins, and poor spin recovery characteristics. During their development cycles, these designs had been subjected to extensive wind tunnel tests to tailor their performance capabilities and to spin tunnel tests to evaluate spinning and spin recovery behavior. However, because their expected operational missions did not require flight at high angles of attack, wind tunnel tests for high-angle-of-attack and stall characteristics were limited or nonexistent.

By the mid-1960s, the number of fighter-attack aircraft and aircrews lost in stall/spin accidents had increased alarmingly, with over 250 aircraft lost by the Air Force, Navy, and Marines from 1965 to 1971.[11] Several major events occurred as a result of this national problem that would impact the role of dynamic free-flight models in fighter development programs. The military at the highest levels became aware of the situation and called for a national effort to improve the high-angle-of-attack stability and control of existing and future aircraft. In responding to this urgency, industry and DOD began to emphasize high-angle-of-attack capabilities, improve and verify design methods, and coordinate testing techniques. Spurred on by the intensity of this national effort, a philosophy emerged in the 1970s that focused on the prevention of loss of control at high angles of attack, rather than recovery from inadvertent spins. In other words, it was rationalized that emphasizing spin and recovery demonstrations during aircraft development programs was working the wrong end of the problem, and that more emphasis should be placed on providing satisfactory handling characteristics during appropriate high-angle-of-attack maneuvers at the edge of the envelope. The military philosophy had changed from avoiding flight at high angles of attack to demanding carefree maneuver capability under those conditions.

With this background, NASA, DOD, and industry formulated an extensive attack on the development of high-angle-of-attack technology, with a key role identified for NASA's unique dynamic model test techniques. The NASA relationships with DOD and industry that emerged in this late 1960s challenge rapidly matured, to the point that nearly all new DOD highly maneuverable aircraft configurations underwent coordinated testing at Langley in wind tunnel free-flight tests, spin tunnel tests, and outdoor drop-model tests.

Previously, testing at Langley had centered on spin tunnel activities, but in the new approach, free-flight testing also became a vital new component of the test programs, with much more emphasis on spin prevention. The model tests were coordinated with, and augmented by, conventional testing in other wind tunnels at Langley and Ames as well as flight research at Dryden. This Government-industry relationship flourished for over 35 years and included high-angle-of-attack free-flight model tests in the Langley Full-Scale Tunnel for the F-4, F-5, F-14, F-15, F-111, YF-16, YF-17, F-16, F-16XL, F/A-18C, X-29, XFV-12A, EA-6B, YF-22, YF-23, X-31, F/A-18E, and F-22. In addition to the fighter test activities, high-angle-of-attack free-flight tests were conducted in response to an Air Force request for the B-1 bomber. Overviews of the test programs are discussed in *Partners in Freedom*, and references for technical details are provided herein.[12]

Langley's free-flight model tests of specific DOD fighter aircraft played a key role in the early phases of aircraft development programs by reducing risk and instilling confidence before full-scale flight tests. All of the tests were witnessed by key industry representatives, who quickly absorbed the results and disseminated them within the design teams. These technical contributions included early identifications of potential dynamic stability and control problems observed during the model flights and aerodynamic data measured with the same model during supporting force and moment tests at the same flight conditions. Industry greatly appreciated this unique source of preflight information, and the free-flight testing technique and NASA's expertise in conducting and interpreting the results were widely sought.

Examples of the technical results and lessons learned in the free-flight tests are noteworthy. In the case of the F-4, the Air Force approached NASA for assistance in defining the major factors causing poor high-angle-of-attack behavior and an alarming increase in loss-of-control accidents. Ironically, the F-4 had been in service for years before aerodynamic data at high angles of attack were measured in tunnel tests at Langley.[13] Operational experience with the F-4 had shown that, as angle of attack was increased at sub-

sonic maneuvering conditions, lightly damped roll oscillations were encountered, which continued to build to large amplitudes on the order of plus or minus 20 degrees. If the pilot forced the aircraft to even higher angles of attack, the F-4 would suddenly exhibit directional divergence, abruptly switching nose for tail, and enter an incipient spin. The departure from controlled flight was also aggravated by unsatisfactory adverse yaw from aileron deflections. When free-flight model tests of the F-4 were conducted in the Full-Scale Tunnel, the model faithfully reproduced all of these phenomena, including visible adverse yaw, large-amplitude wing rock, and a directional divergence that required quickly removing the out-of-control model from the tunnel airstream.

Analysis of supporting force and moment test data measured with the free-flight model provided

Free-flight tests of this model of the F-4 Phantom II in 1970 provided data for analysis of the wing rock and directional divergence exhibited by the configuration at high angles of attack. The test program also included demonstrations of the benefits of two-position fixed leading-edge slats, which were incorporated in later versions of the airplane.

substantial information on the causes and potential cures for the deficient characteristics. As the analysis progressed, military interest was building for a derivative of the baseline F-4 configuration, which would

include wing leading-edge slats to improve maneuverability during dogfights. Extensive testing of the configuration had been conducted during conventional wind tunnel tests by McDonnell-Douglas and Langley in the Langley High-Speed 7- by 10-Ft Tunnel, and free-flight evaluations were requested to assess the impact on dynamic stability and control. Flight tests of the slatted wing configuration in 1970 indicated substantial improvement in high-angle-of-attack characteristics, including reduced wing-rock oscillations and delayed tendency toward directional divergence. When the advanced configuration, known as the F-4E, entered service, these enhanced characteristics were evident.

During the free-flight model support activities for the development of the F-111 aircraft, the model was subjected to preflight conventional force tests in the 12-Foot Low-Speed Tunnel to obtain aerodynamic stability and control data at high angles of attack. Force tests such as these were routinely conducted before models were subject to flight tests in the Full-Scale Tunnel.

Free-flight tests of the General Dynamics F-111 in the Full-Scale Tunnel in 1964 and 1965 uncovered a similar directional divergence problem for certain wing configurations, and the severity of the directional divergence made an impression on visiting pilots who were slated to conduct full-scale flight tests for high angle of attack. In addition to alerting the flight-test organization to the potential problem, the results stimulated the F-111 aircraft designers to develop a stall-inhibitor capability in the flight control system to artificially limit the maximum obtainable angle of attack and thereby avoid the divergence.

Another particularly successful NASA–industry–DOD free-flight model test program was conducted during the competitive Air Force Lightweight Fighter program in the mid-1970s. Free-flight investigations of the high-angle-of-attack dynamic stability and control characteristics of the General Dynamics (now Lockheed Martin) YF-16 and the Northrop (now Northrop Grumman) YF-17 configurations provided critical technical data to each of the industry flight-test teams and for the Air Force evaluation team, which ultimately selected the YF-16 for development. The results

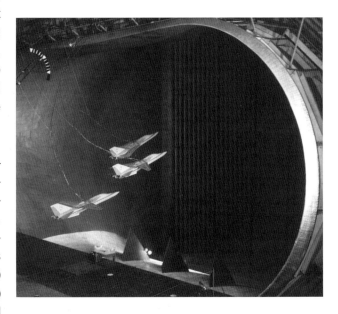

Free-flight tests of the F-111 model in 1965 demonstrated the severity of the directional divergence encountered for certain values of wing sweep at high angle of attack. This photographic sequence shows the model in flight for forward, mid, and aft wing-sweep positions.

of the free-flight test of the YF-16 were especially valuable because they identified a severe nose-slice tendency for the configuration at angles of attack slightly beyond maximum lift. Fortunately, the YF-16 design team at General Dynamics had anticipated the problem based on the previously discussed NASA-developed criterion for prediction of directional divergence. Its analysis showed that an angle-of-attack-limiting system would prohibit the full-scale aircraft from reaching the critical angle of attack yet provide superior maneuverability in air combat.

In a broad cooperative research program with General Dynamics to develop a supersonic-cruise version of the F-16 in the 1980s, Langley researchers used free-flight testing in the Full-Scale Tunnel to identify major stability and control issues at high angles of attack for the cranked wing F-16XL configuration. Results of the model flight tests identified unacceptable pitch-up and directional-divergence instabilities for the original design. After additional joint testing and free-flight evaluations, geometric modifications were made to the F-16XL configuration wing apex area and wing upper surface, which significantly improved its characteristics and resulted in outstanding behavior at high angles of attack. Subsequently, full-scale flight tests verified the highly satisfactory characteristics predicted by the model flight tests.

Free-flight tests of the YF-16 Lightweight Fighter Prototype demonstrated outstanding dynamic stability control characteristics within the design envelope. The results also clearly identified a severe directional divergence, which occurred for high angles of attack beyond achievable values limited by the flight control system.

Configuration modifications to the General Dynamics F-16XL aircraft included geometric shaping of the wing apex area and wing fences. These modifications were derived from highly successful free-flight and force tests conducted in the Full-Scale Tunnel in 1981.

The Grumman X-29 Advanced Technology Demonstrator research airplane provided a challenge for the free-flight model technique in the Full-Scale Tunnel during the early 1980s.[14] The X-29 was designed with a large level of aerodynamic instability in pitch (relaxed static stability) that made the use of artificial stability and control mandatory for meaningful model tests. The magnitude of the instability was by far the highest ever attempted with a free-flight model. Accepting the challenge, the research staff configured the updated digital computer used in free-flight testing to replicate the stabilizing features of the full-scale airplane flight control system. The challenge was met by using an angle-of-attack vane sensor on the model's nose boom for feedback inputs to the computer in an arrangement that worked flawlessly through the flight-test program. With that experience, the free-flight wind tunnel model technique had progressed to its most sophisticated level of capability.

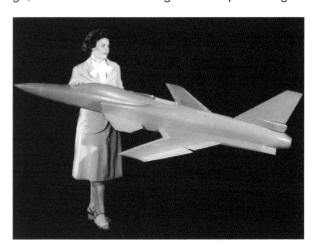

The ability to conduct free-flight model tests of the X-29 research aircraft required a significant upgrade in Langley's testing technique to replicate the artificial stability provided by the flight control system of the full-scale aircraft. High-angle-of-attack flight tests of the model were conducted in 1980.

In 1989, free-flight tests of the Lockheed Martin YF-22 prototype provided what proved to be accurate predictions of the unprecedented high-angle-of-attack capabilities of the full-scale aircraft. After years of coordinated activities in the newly integrated high-angle-of-attack and spin research organizations at Langley, the YF-22 model efforts were especially effective and resulted in a formal letter of recognition and thanks to NASA from the management of Lockheed Martin.[15] After the competitive contract between the YF-22 and YF-23 aircraft was awarded to the YF-22 organization, no request was made for traditional free-flight model tests in the Full-Scale Tunnel. Rather, a typical free-flight-sized model was used in static and dynamic force testing at Langley to support the F-22 development program.

In addition to providing support for specific aircraft development programs, NASA's interest in high-angle-of-attack technology led it to develop a major intercenter research program to advance the state of the art in experimental, computational, and flight-testing methods. The program also addressed emerging technologies such as thrust vectoring and fuselage-forebody flow-control concepts for improved capabilities at high angles of attack. Participation in the NASA program included the Langley, Dryden, Ames, and Lewis (now Glenn) Research Centers. The

Free-flight model tests of the YF-22 prototype in 1989 had demonstrated outstanding stability and control characteristics at high angles of attack, which were subsequently verified in the competitive evaluation flight tests of the aircraft. The follow-on F-22 program did not request a NASA free-flight study, but a conventional force-test program was conducted in the Full-Scale Tunnel in 1992 with a model similar to those used in free-flight testing.

NASA High-Alpha Technology program used the F/A-18 configuration as the focus of its research efforts and included extensive tests with free-flight models. Free-flight model tests in the Full-Scale Tunnel had been conducted at the request of the Navy in support of the F/A-18 development program, and NASA researchers used results from that program and correlation with known full-scale flight characteristics as baseline data for comparison with benefits provided by radical new technologies in the NASA research program.[16]

In the 1980s, two breakthrough concepts emerged that would prove to revolutionize high-angle-of-attack technology. The first revolution was the routine use of advanced digital fly-by-wire flight control systems to provide unprecedented maneuverability at high angles of attack. Such concepts used various combinations of control surfaces to simultaneously enhance aircraft responses to pilot inputs while preventing loss of control and inadvertent spin entry. The second critical technology was the development of thrust-vectoring concepts for enhanced controllability and stability for high-angle-of-attack conditions. Thrust vectoring provided unprecedented levels of control power when the effectiveness of conventional control surfaces markedly deteriorated at high angles of attack. Together with the development of engines that could operate reliably at high angles of attack and sideslip, vectoring offered new options for control. Free-flight models proved to be an effective approach to evaluate and demonstrate the impressive capabilities provided by these technologies.

As interest in thrust vectoring intensified in the 1980s, other Langley organizations designed supersonic advanced configurations using thrust vectoring as the primary control concept. This tailless design of a supersonic fighter using pitch- and yaw-vectoring vanes for control was flown at extreme angles of attack during free-flight tests in the Full-Scale Tunnel.

Free-flight tests in the Full-Scale Tunnel with a large number of fighter configurations demonstrated that thrust vectoring in pitch and yaw nearly eliminated previous constraints on flight at high angles of attack. Control effectiveness in pitch, yaw, and roll remained at a high level for angles of attack approaching 90 degrees, and these powerful control moments were also used with appropriate flight sensors to increase the stability of aircraft at high angles of attack. To evaluate thrust vectoring in flight with a full-scale F/A-18 aircraft with low-cost hardware modifications, NASA designed thrust-vectoring paddles in the engine exhausts of the twin-engine aircraft. After extensive engineering analysis and supporting tests in other facilities, free-flight model tests were conducted to evaluate the potential benefits of the candidate factoring system.

Thrust vectoring provided high levels of control at high angles of attack, at which conventional aerodynamic controls are degraded. In this free-flight test of 1983, pitch and yaw vanes provided control power, and an angle-of-attack vane provided feedback for stability augmentation to the extent that the vertical tails could be removed without effect for flight at extreme angles of attack.

During flights in the Full-Scale Tunnel, the F/A-18 free-flight model powered by compressed-air engine simulators and equipped with vectoring paddles could be precisely controlled at extreme angles of attack. The relative effectiveness of thrust vectoring was impressively demonstrated when the twin vertical tails were removed with no degradation in control at angles of attack greater than 30 degrees. Subsequently, the powerful effects of vectoring were demonstrated in flight with a modified F/A-18 full-scale aircraft known as the High-Alpha Research Vehicle (HARV).[17] In addition to the F/A-18 application, the free-flight technique demonstrated that similar results could be obtained with models of the F-16 and X-29 aircraft.

In addition to showing the benefits of thrust vectoring when retrofitted to existing aircraft configurations, Langley continued its exploratory assessments of applications for high-angle-of-attack conditions with several NASA generic-research models configured without horizontal or vertical tails. With thrust vectoring in pitch and yaw, flight demonstrations of these tailless models provided impressive statements of the unprecedented level of controllability provided by thrust vectoring. Although such tailless configurations would require auxiliary backup control surfaces for emergency engine-out conditions, the test results suggest that future designs (including stealthy configurations with reduced radar signatures) could employ vectoring concepts for operations at extreme angles of attack.

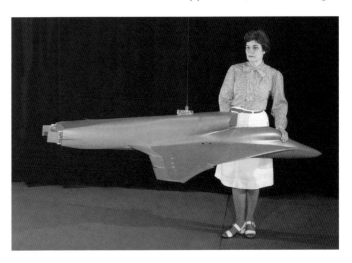

As interest in thrust vectoring intensified in the 1980s, other Langley organizations designed supersonic advanced configurations using thrust vectoring as the primary control concept. This tailless design of a supersonic fighter using pitch- and yaw-vectoring vanes for control was flown at extreme angles of attack during free-flight tests in the Full-Scale Tunnel.

Another valuable contribution of free-flight model flight tests to high-angle-of-attack technology was a demonstration of yaw control effectiveness using deflectable small strakes near the nose tip of the fuselage forebody. The actuated nose strakes for enhanced rolling (ANSER) system, which had been designed to control powerful vortex flows shed off the nose at high angles of attack, produced large yawing moments (orders of magnitude greater than conventional rudders) at extreme angles of attack.[18] The strakes were also subsequently flown on the HARV aircraft at Dryden in convincing demonstrations of enhanced maneuverability.

The free-flight model of the F-18 flies at high angles of attack with thrust-vectoring vanes and deflectable nose strakes. Both concepts were subjects of intense flight-test investigations at Dryden.

Many other free-flight tests were conducted in the Langley Full-Scale Tunnel in investigations of high-angle-of-attack characteristics of high-performance aircraft. Free-flight tests were conducted to evaluate whether dynamic rolling motions would cause significant effects during flights of a full-scale F-106B aircraft equipped with vortex flaps on the wing-leading edge. Satisfactory results from those tests reduced the risk of the flight-test program and helped its assessments. Another major program for Langley was its partnership in the development of the thrust-vectoring X-31 international experimental aircraft, which demonstrated "super-maneuverability" capabilities at high angles of attack.[19]

NASA activities in support of the international X-31 research aircraft program included free-flight model testing in the Full-Scale Tunnel.

With the transfer of the Langley Full-Scale Tunnel in 1995 to Old Dominion University, NASA's last high-angle-of-attack test program in the facility was a series of force tests for a free-flight model of the F/A-18E.

Free-flight tests of the F/A-18E model were conducted in the Langley 14- by 22-Foot Tunnel.

As discussed, NASA ceased its management of the Langley Full-Scale Tunnel in 1995 and turned operations of the facility over to Old Dominion University. The last military high-angle-of-attack activities conducted by NASA in the facility were conventional force tests of a free-flight model of the F/A-18E configuration in 1996. A brief series of free-flight tests was conducted in the Langley 14- by 22-Foot Tunnel under that program.

Summary

In summary, the role of free-flight wind tunnel model testing for high-performance military configurations became vital in providing superior and safe characteristics for high-angle-of-attack conditions. Development programs for nearly every fighter since the F-4 requested NASA support to identify potential problems early in design and to prepare military flight-test organizations for subsequent full-scale flight experiences. With the closure of the Langley Full-Scale Tunnel and the dramatic reduction in highly maneuverable aircraft programs, the wind tunnel testing technique will face new challenges for future applications.

CHAPTER 6:

SPINNING AND SPIN RECOVERY

Background

The most extensive use of dynamically scaled models by the NACA and NASA in aeronautical programs has been with spinning. Providing satisfactory spin and recovery characteristics has demanded the development and validation of satisfactory prediction techniques for this primary safety-of-flight concern. The complexity of the aerodynamic environment during dynamic spinning motions at poststall angles of attack continues to be challenging for theoretical techniques or methods other than dynamic model tests. After the design, development, and initial operations of free-spinning tunnels by the NACA, nearly 600 spin tunnel entries have been conducted at Langley to assess characteristics of military and civil aircraft. The scope of investigations typically includes assessments of the effects of a large number of variations in aircraft center-of-gravity locations, mass distribution, initial control settings, and recovery techniques. In the case of military configurations with numerous combinations of external stores, test programs in the Spin Tunnel may last for months during exhaustive assessments of the effects of the foregoing variables.[1]

Some of the major results noted from spin tunnel tests include the types of spins (spin modes) exhibited by a configuration, the most effective control combination for recovery, and assessments of the effectiveness of auxiliary spin recovery devices such as parachutes or rockets. The spins exhibited by aircraft include a variety of motions involving steady, unsteady, or oscillatory rotary movement. The classic steady spin mode involves a smooth rotation of the aircraft at a relatively constant angle of attack about a vertical spin axis. If the angle of attack exhibited during the spin is slightly greater than the stall angle of attack, but less than about 45 degrees, the spin is referred to as a steep steady spin, and the spin axis is well forward of the nose of the airplane. On the other hand, if the angle of attack of the spin approaches 90 degrees, the fuselage will approach a horizontal attitude, and the spin is referred to as a steady flat spin about a spin axis near the center of gravity of the airplane. Oscillatory spins involve periodic fluctuations in aircraft pitch, roll, and yaw rates. Steep and moderate spins may be oscillatory; however, flat spins are usually steady. Unsteady poststall modes may also be encountered in which wildly fluctuating, nonperiodic motions occur. Some configurations may exhibit a poststall condition, in which the aircraft enters a deep stall trimmed condition with no significant rotation. Finally, some unconventional aircraft may tumble end over end with no

Analysis of aerodynamic phenomena in spinning began with tests in the Langley 5-Foot Vertical Tunnel in the 1920s. A model of the BT-9 airplane is undergoing static force and moment measurements to obtain data for theoretical studies. The flow in the tunnel was downward, and free-spinning tests were not possible. For some tests, the model was attached to a rotating spin balance, which simulated spinning motions. The main purpose of the vertical orientation of the tunnel was to simplify the balance of rotating masses for dynamic tests.

means to recover. This section of the document will discuss spins, and later sections will address the use of dynamic models to study poststall motions such as tumbling and deep stalls.

Qualitatively, recovery from the various spin modes is dependent on the type of spins exhibited, the mass distribution of the aircraft, and the sequence of controls applied. Recovering from the steep steady spin tends to be easy, because the orientation of the aircraft control surfaces to the free stream enables at least a portion of the control effectiveness to be retained. In contrast, during a flat spin, the control surfaces are oriented so as to provide little recovery moment, especially a rudder on a conventional vertical tail. In addition to the ineffectiveness of controls for recovery from the flat spin, the rotation of the aircraft about a near-vertical axis near its center of gravity results in high centrifugal forces at the cockpit for configurations with long fuselages. In many cases, the negative ("eyeballs out") g-loads will be so high as to incapacitate the crewmembers and prevent them from escaping from the aircraft.

As a result of the accumulation of extensive experience in spin tunnel testing and the correlation of results with full-scale results, Langley researchers have evolved a methodical approach in the use of dynamic models for investigations of spin and recovery characteristics. As will be discussed, this experience base also involves lessons learned regarding potential Reynolds number effects and the necessary modifications to testing techniques to minimize such phenomena.

Military Configurations

After the operational readiness of the Langley 15-Foot Spin Tunnel in 1935, initial testing centered on establishing correlation with full-scale flight-test results of spinning behavior for the XN2Y-1 and F4B-2 biplanes.[2] Critical comparisons of earlier results obtained on small-scale models from the Langley 5-Foot Vertical Tunnel and full-scale flight tests indicated considerable scale effects on aerodynamic characteristics; therefore, calibration tests in the new tunnel were deemed imperative. The results of the tests for the two biplane models were encouraging in terms of the nature of recovery characteristics and served to inspire confidence in the testing technique and promote future tests. During those prewar years, the NACA staff was afforded time to conduct fundamental research studies and to make general conclusions for emerging monoplane designs. Systematic series of investigations were conducted in which, for example, models were tested for combinations of eight different wings and three different tails.[3] Other investigations of tunnel-to-flight correlations occurred, including comparison of results for the BT-9 monoplane trainer.

As experience with spin tunnel testing increased, researchers began to observe more troublesome differences between results obtained in flight and in the tunnel. The effects of Reynolds number, model accuracies, control-surface rigging of full-scale aircraft, propeller slipstream effects not present during unpowered model tests, and other factors became appreciated to the point that a general philosophy began to emerge for which model tests were viewed as good predictors of full-scale characteristics, but examples of poor correlation required even more correlation studies and a conservative interpretation of model results. Critics of small-scale model testing did not accept a growing philosophy that spin predictions were an "art" based on extensive testing to determine the relative sensitivity of results to configuration variables, model damage, and testing technique. Nonetheless, pressure mounted to arrive at design guidelines for satisfactory spin recovery characteristics. An empirical criterion based on the projected side area and mass distribution of the airplane was derived in England, and the Langley staff proposed a design criterion in 1939 based solely on the geometry of aircraft tail surfaces. Known as the tail-damping power factor (TDPF), it was touted as a rapid estimation method for determining whether a new design was likely to comply with the minimum requirements for safety in spinning.[4]

The beginning of World War II and the introduction of a Langley 20-Foot Spin Tunnel in 1941 resulted in a tremendous demand for spinning tests of high-priority military aircraft. The workload of the staff increased dramatically, and a tremendous amount of data was gathered for a large number of configurations. Military requests for spin tunnel tests filled all available tunnel test times, leaving no time for general research. At the same time, configurations were tested with radical differences in geometry and mass distribution.

Tailless aircraft with their masses distributed in a primarily spanwise direction were introduced, along with twin-engine bombers and other unconventional designs with moderately swept wings and canards.

After the breathtaking demands of the war were met, the NACA staff turned its interests back toward improving the state of the art for spin tunnel testing and developing spin recovery design criteria. Once again, the results of spin tunnel model tests were examined to arrive at design guidelines for future aircraft.[5] The earlier TDPF criterion was reexamined in light of additional spin tunnel results gathered during the war and found to be deficient based on the more recent results. In addition, large effects of the relative mass distribution along the wings and fuselage observed in more recent tests had not been incorporated. In view of these shortcomings, an effort was undertaken to develop a revised empirical criterion based on results of tests of about 100 airplane designs in the 15-foot and 20-foot spin tunnels.

The procedure for updating the TDPF criterion involved plotting a parameter based on the geometry of the tail as a function of the mass distribution for three values of the density of the aircraft relative to the atmosphere. Results obtained in tunnel tests were then located on the plot, and boundaries were drawn to separate regions where spin recoveries were satisfactory and those where recoveries were unsatisfactory. The configurations used for the analysis included biplanes, monoplanes, land-planes and seaplanes, and single- and twin-engine designs. The resulting criterion was intended to be a qualitative guide in the preliminary design of aircraft at the time by offering a reasonable probability of satisfactory recovery by reversal of rudder and elevator. In lieu of further research, industry used this criterion from the late 1940s for design. As will be discussed, the criterion was revisited during a NASA spin research program for general-aviation aircraft in the 1970s.

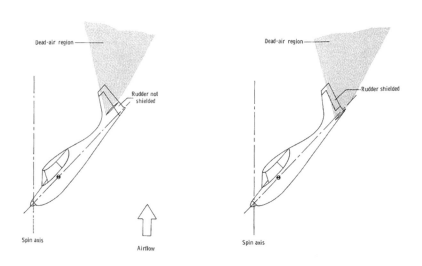

Geometric properties of the tail surfaces of aircraft were major factors in the derivation of the NACA–developed, tail-damping power factor parameter. The aircraft on the left has a better tail design for spinning because the rudder is not shielded by the wake of the horizontal tail and side area is available beneath the horizontal for spin damping.

During the hectic years of World War II, the technical output of spin tunnel testing was necessarily focused on experimental results obtained with the free-spinning models. It was realized, of course, that advances in the state of the art for spin predictions would require a more complete understanding of aerodynamic phenomena experienced during spins. As discussed, aerodynamic measurements for rotating wings and aircraft configurations had been made in the 5-Foot Vertical Tunnel. However, the advent of the Langley 15-foot and 20-foot spin tunnels discontinued that capability in favor of visual observations of the free-spinning models with six degrees of freedom. Immediately after the war, the Langley staff was able to respond to the need for a better understanding of aerodynamic data with the design and operational introduction of a new "rotary balance" in the 20-Foot Spin Tunnel.[6] However, the relatively small free-spinning models were not directly adaptable for testing with the rotary balance, requiring the construction of a new, larger model for aerodynamic force and moment measurements. The use of this test apparatus in conjunction with free-spinning tests significantly improved the understanding of aerodynamic interactions during spins.

In the 1950s, advances in aircraft performance provided by the introduction of jet propulsion resulted in radical changes in aircraft configurations, creating challenges for spin technology. Military fighters no longer resembled the aircraft of World War II, as the introduction of swept wings and long, pointed fuselages became common. Suddenly, certain factors such as mass distribution became more important, and airflow around the unconventional, long fuselage shapes during spins dominated the spin behavior of some configurations. At the same time, fighter aircraft became larger and heavier, resulting in much higher masses relative to the atmospheric density, especially at high altitudes. Once again, the industry and military services turned to the 20-Foot Spin Tunnel at Langley for guidance in the design and testing of these aircraft. As it proceeded to honor requests for spin tunnel testing, the Langley staff realized that most of the technology from the World War II era would no longer apply and that a new experience base would be needed.

A general-aviation model is mounted on a rotary-balance test apparatus in the spin tunnel for measurements of aerodynamic forces and moments during simulated spins. An electrical strain-gauge balance is contained within the model, and the model's attitude relative to the vertically rising airstream is controlled remotely.

Free-spinning model tests of these aircraft shapes quickly identified new phenomena that would shape spin and recovery analysis for years. Slender swept wing models exhibited the normal steep or flat spin modes shown by earlier configurations, but they also displayed oscillatory spin modes that were expected to be disorienting to pilots. The oscillatory spins would also require new recovery techniques. Because of the distribution of mass along the fuselage, the axis of least inertial resistance had become the roll axis, and the use of ailerons became a powerful mechanism for spin recovery.[7] Spin tunnel tests of models of the Bell X-5 variable-sweep research aircraft (at aft sweep) and the slender, pointed-nose Douglas X-3 Stiletto were among the first to document the new behavior.

During the late 1930s, British researchers had noted considerable effects of Reynolds number on the aerodynamic behavior of certain fuselage cross-sectional shapes during spins. In particular, the aft-fuselage components of aircraft configurations having square or rectangular aft-fuselage, cross-sectional

Radical new aircraft configurations of the 1950s had significant impacts on the state of the art in spinning by introducing tailless designs and extreme variations in mass distribution along the fuselage. The slender X-3 research aircraft, center, and the X-5 variable-sweep design, upper right, were the first to demonstrate the effectiveness of roll control for spin recovery during tests of free-spinning models at Langley.

shapes with certain types of rounded corners were discovered to produce pro-spin yawing moments at low Reynolds number representative of spin tunnel tests but anti-spin moments for full-scale values of Reynolds number.[8] As expected, such characteristics had huge effects on the quality of correlation between model predictions and full-scale tests.

In the mid-1950s, the NACA encountered similar effects of cross-sectional shape, but in this case, the phenomenon involved the long fuselage forebodies being introduced at the time. This experience led to one of the more important lessons learned in the use of free-spinning models for spin predictions. One project stands out as a key experience regarding this topic. As part of the ongoing military requests for NACA support of aircraft development programs, the Navy asked Langley to conduct spin tunnel tests of a model of its new Chance Vought XF8U-1 Crusader fighter in 1955. The results of spin tunnel tests of a 1/25-scale model indicated that the airplane would exhibit 2 spin modes.[9] The first mode would be a potentially

Spin tunnel tests of the Chance Vought F8U fighter resulted in one of the more important lessons learned in dynamic free-flight model testing. Unrealistic pro-spin moments produced by the aerodynamic behavior of the cross-sectional shape of the forward fuselage resulted from the low Reynolds number of the free-spinning test. Subsequently, configurations were checked for possible sensitivity to such effects before tests and artificial devices used to minimize the effect.

dangerous fast flat spin at an angle of attack of approximately 87 degrees, from which recoveries were unsatisfactory or unobtainable. The second spin was much steeper with a lower rate of rotation, and recoveries would probably be satisfactory.

As the spin tunnel results were analyzed, Chance Vought engineers directed their focus to identifying factors that were responsible for the flat spin exhibited by the model. The scope of activities stimulated by the XF8U-1 experience included, in addition to extended spin tunnel tests, one-degree-of-freedom autorotation tests of a model of the XF8U-1 configuration in the Chance Vought Low Speed Tunnel and a NACA wind tunnel research program that measured the aerodynamic sensitivity of a range of two-dimensional, noncircular cylinders to Reynolds number.[10] The wind tunnel tests were designed and conducted to include variations in Reynolds number from the low values associated with spin tunnel testing to higher values more representative of flight.

With results from the wind tunnel studies in hand, researchers were able to identify an adverse effect of Reynolds number on the forward fuselage shape of the XF8U-1 such that at the relatively low values of Reynolds number for the spin tunnel tests (about 90,000, based on fuselage forebody depth), the spin model exhibited a powerful pro-spin yawing moment dominated by the forebody. The pro-spin moment caused an autorotative spinning tendency resulting in the fast flat spin observed in the spin tunnel tests. As Reynolds number in the conventional tunnel tests was increased to values approaching 300,000, however, the moments produced by the forward fuselage became anti-spin and remained so for higher values of Reynolds number. The researchers had identified the importance of cross-sectional shapes of modern aircraft—particularly those with long forebodies—on spin characteristics and the possibility of erroneous spin tunnel predictions. When the full-scale spin tests were conducted, the XF8U-1 airplane exhibited only the steeper spin mode, and the fast flat spin predicted by the spin model was not encountered.

The mechanism of the Reynolds number effect experienced in the XF8U-1 project is depicted in the accompanying sketch, which shows a top view of an airplane descending in a vertical spin to the right at a high angle of attack, representing a flat spin. The arrows along the nose indicate the relative magnitude and direction of the sideward velocities induced along the fuselage because of the spinning rotation. The sketch on the right is a head-on view that illustrates the local sideslip angle created at a representative nose location (section A-A) because of the spin rotation. The airplane's near-vertical rate of descent and the sideward velocity at the nose combine to produce a positive sideslip angle at the nose. If, for the right-spin condition, nose-right aerodynamic yawing moments are produced by the local sideslip angle, the resulting moment will tend to increase the spin rate. If, on the other hand, nose-left moments are produced, the aerodynamic moment will tend to resist the spin.

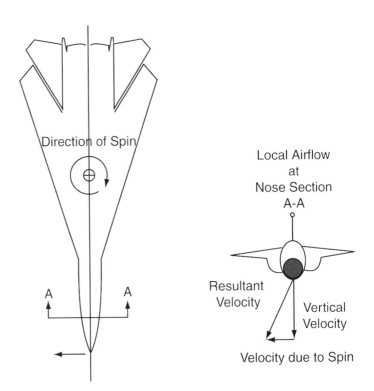

Local flow angularity at a representative fuselage forebody location caused by the spin rate and the vertical velocity of airplane. Shown in top and front views during a right spin, the section A-A experiences a local sideslip to the right.

The type of moment produced by certain nose shapes can be sensitive to Reynolds number. For example, the accompanying sketch shows the variation of yawing moment produced by a square cross-sectional shape with rounded corners subjected to variations in Reynolds number at an angle of sideslip of 5 degrees at an angle of attack of 80 degrees, representative of a flat spin.

Two regions of Reynolds number are of interest: low values, typical of tests in the Spin Tunnel, and high values, representative of full-scale conditions. The values of yawing moment are pro-spin for the model test condition but become anti-spin for the full-scale flight condition at high Reynolds number. This trend in aerodynamic behavior is caused by movement of the flow separation areas on the fuselage cross section with variations in Reynolds number.

During decades of experience in correlating model and full-scale airplane flight results after the XF8U-1 project, Langley's spin tunnel personnel developed expertise in the anticipation of potential Reynolds number effects, such as the forebody effect, and in the art of developing methods to geometrically modify models to minimize undesirable effects from the phenomenon. In this approach, cross-sectional shapes of aircraft are examined before models are constructed, and if the design is similar to those known to exhibit scale effects, static wind tunnel tests are conducted for a range of Reynolds number to determine if artificial devices such as nose-mounted strakes can be used to alter the flow separation on the

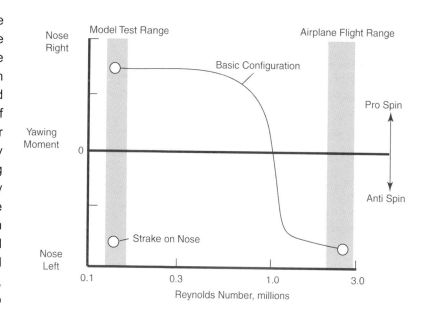

Adverse effect of Reynolds number on the forebody of a fighter model as measured in wind tunnel tests. At relatively low values of Reynolds number, the model displays propelling yawing moments caused by flow separation phenomena at that condition, whereas at full-scale Reynolds numbers, the moment is anti-spin because of changes in flow separation. Also shown is the favorable effect of adding nose strakes.

nose at low Reynolds number and more accurately simulate full-scale conditions.[11] As shown by the data point at model values of Reynolds number in the sketch, the addition of a small fuselage strake to some shapes can produce model aerodynamics comparable to those of the airplane. When strakes are applied to spin tunnel models to correct for scale effects, they are only used for the range of angles of attack where they are meant to be effective, and the model is hand-launched into the tunnel at those angles of attack.

Interestingly, after testing had been completed and lessons had been learned for the XF8U-1, a derivative of the aircraft known as the F8U-1P designed for reconnaissance missions required testing in the Langley Spin Tunnel. This version of the airplane included a modification to the lower fuselage forebody cross section to accommodate a camera installation. The forebody modification resulted in a cross-sectional shape with less rounded corners for the forward fuselage. Based on the earlier XF8U-1 experiences, precursor static wind tunnel tests of the new configuration were conducted in the Langley 300-mph 7- by 10-Foot Tunnel to assess the impact of Reynolds number variations for the new configuration. The results indicated that the cross-sectional shape of the F8U-1P would produce a propelling pro-spin yawing moment at angles of attack of 70 degrees or greater for both the spin tunnel model and the full-scale airplane. In this case, the fast flat spin expected in the tunnel tests was verified, and the flight program could approach spin tests with confidence that the flat spin would be encountered.[12]

In addition to the XF8U-1, it was necessary to apply scale-correction fuselage strakes to the spin tunnel models of the Northrop F-5A and F-5E fighters, the Northrop YF-17 lightweight fighter prototype, and the

Fairchild A-10 attack aircraft to avoid erroneous predictions because of fuselage forebody effects. In the case of the X-29, a specific study of the effects of forebody devices for correcting low Reynolds number effects was conducted in detail.[13]

The introduction of fuselage shapes with long pointed noses also resulted in another unexpected phenomenon that greatly influenced spin and recovery characteristics. It was noted that spin tunnel models with pointed noses exhibited spin and recovery characteristics that were sensitive to the physical condition of the nose tip. Sometimes, seemingly unimportant dents or deformation of the extreme nose tip greatly affected free-spinning results. Some models spun readily in one direction and not at all in the other, whereas later in the test program, the direction in which the model would spin reversed, a trend that continued through the tunnel test program as a model suffered wear and tear during testing. To provide insight to the mechanisms involved, the NACA and NASA conducted conventional wind tunnel force and moment testing, which defined the critical factor to be large asymmetric yawing moments produced on the forebody as a result of slight imperfections near the nose tip.[14] The magnitude of the asymmetric moment could be large, reaching values many times larger than those provided by corrective controls.

The powerful effects of asymmetric moments were observed in free-spinning model and full-scale flight tests for the X-3 research airplane, F-104 fighter, and F-111 fighter-bomber.[15] Inspired by the results being obtained in the Spin Tunnel, researchers at the Ames Research Center conducted informative aerodynamic testing of generic pointed nose shapes over a range of Mach number and Reynolds number in the Ames 12-Foot Pressure Tunnel to provide basic design information.[16]

As experience with nose-induced asymmetric yawing moments was accumulated, researchers explored the use of flow-control devices to enable spin recovery or even prevent spins. The progress of the work began with free-spinning studies using several models to demonstrate that rotating a small section of the nose tip could control the direction of the spin. Next, the researchers conducted free-spin, force, and moment testing to provide guidelines for the use of auxiliary devices such as long thin strakes along the forebody or foldout canards for spin recovery. Model spin tests were even conducted to demonstrate that actively rotating a pointed nose during the spin could result in spin recovery.[17] The expertise and experiences accumulated by the research staff of the Spin Tunnel regarding the impact of long pointed fuselage shapes ultimately provided major inputs for several full-scale aircraft development programs, including the F-104, F-111, X-29, and X-31 programs.

External stores have been found to have large effects on spin and recovery, especially for asymmetric loadings, in which stores are located asymmetrically along the wing, resulting in a lateral displacement of the center of gravity of the configuration. For example, some aircraft may not spin in the direction of the "heavy" wing but will spin fast and flat into the "light" wing. In most cases, model tests in which the shapes of the external stores were replaced with equivalent weight ballast indicated that the effects of asymmetric loadings were primarily due to a mass effect, with little or no aerodynamic effect detected. However, large stores such as fuel tanks were found, on occasion, to have unexpected effects because of aerodynamic characteristics of the component. During the aircraft development, spin characteristics of high-performance military aircraft must be assessed for all loadings proposed, including symmetric and asymmetric configurations. Spin tunnel tests can therefore be extensive for some aircraft, especially those with variable-sweep

Free-spinning tests are used to determine the effects of external stores on the spin and recovery characteristics of military aircraft. The photograph shows a model of the F/A-18F during store effects testing after the deployment of the emergency spin recovery parachute.

wing capabilities. Testing of the General Dynamics F-111, for example, required months of test time to determine spin and recovery characteristics for all potential conditions of wing-sweep angles, center-of-gravity positions, and symmetric and asymmetric store loadings.

The effects of other configuration and flight variables are also considered in planning free-spinning model tests. For example, the direct and indirect effects of propulsive power on spin and recovery characteristics have been addressed. Although the policy for recovery from spins in propeller-driven airplanes called for retarding the throttle as soon as possible during the spin because of possible adverse effects, in some cases, the pilots recovered from otherwise uncontrollable spins by applying full power. Such results may have been due to increased effectiveness of the rudder and elevator surfaces in the propeller slipstream. With the advent of jet engines, however, direct impingement of jet exhaust on control surfaces was unlikely.

However, new concern arose over possible effects of gyroscopic moments produced by the rotating parts of jet engines.

To assess the potential impacts of gyroscopic effects on spins and recovery, Langley conducted free-spinning model tests in which rotating flywheels powered by model airplane engines were used to simulate the expected gyroscopic behavior.[18] Results of exploratory tests for a representative straight wing attack airplane model with a flywheel designed to simulate a range of gyroscopic moments showed effects on spin and recovery with different results depended on spin direction. In particular, the rotating flywheel changed spin recovery from satisfactory to unsatisfactory for right spins, but for left spins, the satisfactory recovery characteristics for the basic airplane were not altered. One of Langley's experiences with large, engine-generated gyroscopic effects occurred during spin tunnel evaluations of the Ryan X-13 tailsitter VTOL, which consisted of a small airframe and a large turbojet engine. Today's jet fighters do not, in general, exhibit large gyroscopic effects in spins, and the simulation of engine rotating parts is usually not included in the test program.

In addition to providing predictions of spin and recovery characteristics for the Nation's military aircraft, free-spinning models are used to develop key technologies required for emergency spin recovery devices such as parachutes, rockets, and extensible surfaces such as canards and strakes. Such devices are typically carried by development aircraft specifically targeted for high angle of attack and spin demonstration tests. In the Spin Tunnel, miniature versions of the devices are deployed during spins to evaluate their effectiveness in terminating the spin motion.

The use of tail-mounted parachutes for emergency spin recovery has been common from the earliest days of flight to the present day. Properly designed and deployed parachutes have proven relatively reliable spin recovery devices, always providing an anti-spin moment, regardless of the orientation of the aircraft or the disorientation or confusion of the pilot. Almost every military aircraft spin program conducted in the Spin Tunnel includes a parachute investigation. Free-spinning model tests are used to determine the critical geometric variables for parachute systems. Paramount among these variables is the minimum size of parachute required for recovery from the most dangerous spin modes. As would be expected, the size of the parachute is constrained by issues regarding system weight and the opening shock loads transmitted to the rear of the aircraft. In addition to parachute size, the length of parachute riser (attachment) lines and the attachment point location on the rear of the aircraft are also critical design parameters.

The importance of parachute riser line length can be critical to the inflation and effectiveness of the parachute for spin recovery. Results of free-spin tests of hundreds of models in the Spin Tunnel have shown that if the riser length is too short, the parachute will be immersed in the low-energy wake of the spinning airplane and will not inflate. On the other hand, if the towline length is too long, the parachute will inflate but will drift inward and align itself with the axis of rotation, thereby providing no anti-spin contribution. The design and operational implementation of emergency spin recovery parachutes is a stringent process that begins with spin tunnel tests and proceeds through the design and qualification of the parachute system, including the deployment and release mechanisms. By participation in each of these segments of the process, Langley researchers have amassed tremendous knowledge regarding parachute systems and are called upon frequently by the aviation community.[19]

Other auxiliary spin recovery devices have been explored using dynamically scaled models. Spin tunnel studies have been conducted on the use of spin recovery rockets mounted at the wingtips of model aircraft, including the XF8U-1, T-28, A-5A, and OV-10.[20] Deviation from traditional parachute installations is seriously considered for some configurations in which the carriage and deployment of a parachute would be difficult. For example, the North American OV-10 twin-boom configuration incorporated a unique tail design not suitable for emergency parachute systems.[21] The results of the rocket tests for the various aircraft models defined the impulse (magnitude and duration) of the rocket thrust required for satisfactory recovery and the relative thrust angle required. Analytical studies of the effectiveness of rockets for spin recovery have also been made.[22]

Other studies have included the evaluation of foldout canard surfaces for spin recovery. The most well-known application of this concept was for the F-14 high-angle-of-attack/spin test airplane. Spin tunnel testing of the F-14 configuration indicated the possibility of a fast flat spin with no recovery—even using a combination of conventional aerodynamic controls and a reasonably sized parachute. Free-spinning model tests demonstrated that satisfactory recoveries from the flat spin could be achieved if the maximum allowable parachute was combined with deployable small canard surfaces mounted on the fuselage forebody. A full-scale F-14 test airplane was configured with the parachute and canards for joint NASA–Grumman–Navy high-angle-of-attack flight-testing at the Dryden Flight Research Center.[23]

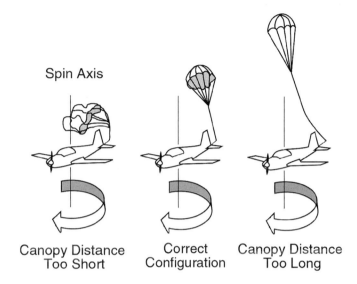

Critical geometrical properties of an emergency spin parachute system include the parachute type and diameter, the riser line lengths, and the attachment point. The sketches illustrate satisfactory and unsatisfactory installations.

The F-14 used in the NASA–Grumman–Navy high-angle-of-attack program at NASA Dryden was equipped with an emergency spin recovery parachute system and foldout canards on the forward fuselage.

General-Aviation Configurations

The dramatic changes in aircraft configurations after World War II required almost complete commitment of the Spin Tunnel to development programs for the military, resulting in stagnation of research for light personal-owner-type aircraft. In subsequent years, designers had to rely on the database and design guidelines that had been developed based on experiences during the war. Unfortunately, stall/spin accidents in the early 1970s in the general-aviation community increased at an alarming rate, with over 28 percent of fatal accidents attributed to stall/spin. Even more troublesome, on several occasions, aircraft that had been designed according to the tail-damping power factor criterion had exhibited unsatisfactory recovery characteristics, and the introduction of features such as advanced general-aviation airfoils resulted in concern over the adequacy and state of the database for the general-aviation community.

Finally, in the early 1970s, the pressure of new military aircraft development programs eased, permitting NASA to embark on studies related to general-aviation aircraft. The scope of research focused on the use of radio-control and spin tunnel models to assess the impact of design features on spin and recovery characteristics, and to develop testing techniques that could be used by the industry. The program was carefully planned and included the acquisition of full-scale aircraft that were modified for spin tests to produce data for correlation with model results.[24]

One of the key objectives of the program was to evaluate the impact of tail design on spin characteristics with alternate tail configurations specifically designed to produce variability in the TDPF parameter. Models were constructed with four tails used in individual test programs, and results were compared with predictions based on the tail design criteria. The range of tails tested included conventional cruciform configurations and low horizontal tail locations and a T-tail configuration.

Model tests of low wing and high wing general-aviation configurations were undertaken in the Spin Tunnel to evaluate the tail design criterion for more modern airplanes and the effects of other geometric features. Results of the testing indicated that tail configuration had a large influence on spin and recovery characteristics, but many other geometric features also influenced the characteristics, including fuselage cross-sectional shape. In addition, seemingly small configuration features, such as wing fillets at the wing trailing-edge juncture with the fuselage, had large effects. The existing TDPF criterion for light airplanes did not correctly predict the spin recovery characteristics of models for some conditions, especially when ailerons were deflected. NASA's report to the industry stressed that, based on these results, TDPF should not be used to predict spin recovery characteristics. However, the criterion did provide a recommended approach to design of the airplane's tail for enhanced spin recovery behavior.[25]

The general-aviation spin research program included extensive correlations of spin tunnel tests with full-scale spin tests for research aircraft.[26] For a low wing research configuration that served as the workhorse of the program, the correlation of results indicated that the spin tunnel model exhibited all the spin modes that were obtained with the aircraft. Although a moderately steep spin mode was accurately predicted, a fast flat spin predicted by the model was at a lower angle of attack (difference of about 10 degrees) and spin rate than those experienced in flight for the airplane.

As part of its research program, NASA continued to provide information on the design of emergency spin recovery parachute systems.[27] Parachute diameters and riser line lengths were sized based on free-spinning model results for high and low wing configurations and a variety of tail configurations. Additionally, guidelines for the design and implementation of the mechanical systems required for parachute deployment (such as mechanical jaws and pyrotechnic deployment) and release were documented.

During its general-aviation stall/spin program, NASA also encouraged industry to use its spin tunnel facility on a fee-paying basis. Several industry teams proceeded to use the opportunity to conduct proprietary tests for configurations in the tunnel. For example, the Beech Aircraft Corporation sponsored the first fee-paid test in the Langley Spin Tunnel for free-spinning model tests of its Model 77 "Skipper" trainer.[28] In such proprietary tests, the industry provided personnel for joint participation in the testing experience.

Skewed Wing Configuration

One of the more unusual aircraft configurations tested in the Spin Tunnel was the skewed wing concept pursued for potential drag reduction at supersonic speeds. Supporting tests of the NASA AD-1 skewed wing research aircraft at Langley consisted of spin tunnel studies of the spin and spin recovery characteristics of a dynamically scaled model of the airplane.[29] As these tests were initiated, researchers pondered over whether the characteristics of the asymmetric swept wing AD-1 configuration would be unusual. Results of the spin tunnel tests indeed showed unusual behavior in that the AD-1 model would not spin in the direction of the sweptback wing panel. For example, no spin could be obtained to the left if the right wing panel were swept forward, whereas spins were readily obtained in the other direction.

Spin tunnel tests of a model of the NASA AD-1 skewed wing research aircraft showed unusual results. The model would not spin in the direction of the sweptback wing panel. As shown here from a ground view, the airplane would not have spun to the left.

Summary

In summary, the application of dynamically scaled models to spinning by the NACA and NASA has provided the military and civil aviation communities with a national asset and a valuable tool to provide guidance and understanding in a critical safety-of-flight area. Data obtained in the model tests have influenced recommended recovery procedures to be used by pilots during developmental spin tests and operational aircraft. With well over 600 configurations tested in the spin tunnel facility, model tests have become a routine segment of aircraft development programs.

CHAPTER 7:

SPIN ENTRY AND POSTSTALL MOTIONS

Spin Entry

Wind tunnel free-flight tests such as those conducted in the Langley Full-Scale Tunnel are used specifically to provide information on flight characteristics of aircraft configurations for angles of attack up to and including the stall/departure. If the model becomes unstable and departs from controlled flight at high angles of attack, the test is terminated by quickly withdrawing the model from the test section with a safety cable, and the severity of the departure and the effectiveness of recovery controls for the out-of-control condition cannot be evaluated. At the other end of the stall/spin spectrum, spin tunnel testing begins with the model essentially in the fully developed spin condition. In spin tunnel testing, the free-spinning model is hand-launched with prerotation into a spin without regard for whether the aircraft can reach the angles of attack and rotation rates from conventional flight.

Experience has shown, however, that many aircraft may exhibit potentially dangerous flat spins in spin tunnel tests, but some of the configurations cannot be flown into the flat spin from conventional flight because of control limitations, aerodynamic factors, or inertial properties. In addition, some aircraft enter a developed spin quickly after a departure from controlled flight, while others must be forced into the spin with prolonged or unrealistic control inputs. It is difficult, therefore, to determine the spin susceptibility of a specific design based only on the tunnel test techniques. In view of this significant void of information, the NACA and NASA developed two additional outdoor free-flight model test techniques to evaluate spin entry and poststall motions.

The helicopter drop-model technique has been used since the early 1950s to evaluate the spin entry behavior of large unpowered models of military aircraft. The objective of these tests has been to evaluate the relative spin resistance of military configurations after various combinations of control inputs and the effects of timing of recovery control inputs after departures. Another testing technique to evaluate spin resistance and out-of-control recovery procedures consists of spin entry evaluations of remotely controlled powered models that take off from prepared runways and fly to the test condition. NASA conducted studies in its research program for general-aviation aircraft using this technique with propeller-driven general-aviation configurations. The objectives of the tests included correlation with spin tunnel and full-scale results, evaluations of the effects of configuration modifications to enhance spin resistance, development of testing technique procedures, and relatively low-cost instrumentation that could be used by industry design teams within the general-aviation community. The latest powered-model testing technique used by NASA involves sophisticated powered models of jet-transport-type configurations for studies of control strategies for recovery of transports after large attitude upset conditions.

Historically, the first use of free-flight models by the NACA and NASA for studies of the incipient spin consisted of hand-launched models that were released with pro-spin controls at an elevated location in a large airship hangar at Langley in the 1920s. Along with British researchers, the NACA experts recognized that the technique was tedious and time-consuming, to the point that testing this approach was abandoned during World War II. After the war, however, the sudden influx of inexperienced private pilots resulted in a dramatic increase in fatal stall/spin accidents that were characterized by inadvertent spin entry at low altitudes and ground impact before the developed spin could occur. At Langley, a lull in spin testing demands by the military allowed for a reexamination of spin entry testing techniques, and a testing technique was developed based on catapult launches of models in a large building that had previously housed the 15-Foot Spin Tunnel.[1]

The first use of the NACA catapult technique included an assessment of the testing efficiency and repeatability of incipient spin motions of a representative personal-owner airplane. The tests were performed inside a building about 70-feet square and 60-feet high, with the catapult launching apparatus 55 feet above a recovery net. An elastic cord was used to catapult the model along a short track at a prescribed high angle of attack for launch, with the elevator control set to pitch the model from the launching angle of attack up to the stall, and the rudder was set to initiate a yawing motion. Data obtained from the tests were based on motion-picture records. The results of this initial study showed that the unpowered model motions were generally repeatable and symmetrical to the right and to the left. However, two major concerns surfaced after the assessment. The first concern was regarding the inefficiency and brief duration of the tests, and the second concern was over the potential existence of scale effects for the small model (wingspan of about 33 inches) used in the investigation. This early experience did not inspire follow-on activities, and the development of more feasible testing techniques for studies of the incipient spin had to await other venues.

Military Configurations

In the late 1950s, military issues once again began to shape testing techniques at Langley. New fighter aircraft configurations began to show potentially dangerous spin modes in spin tunnel tests and operational service, stimulating interest on the possibility of terminating the spin while the aircraft motions were still in the incipient phase. This emphasis was appropriate, because conventional recovery controls were, in many cases, found to be ineffective in terminating a fully developed spin, whereas recovery controls were effective in terminating the incipient spin motions if applied early after loss of control. At that time, military fighters continued to grow in size and weight, and the catapult facility at Langley was grossly inadequate for conducting such investigations. Industry had become concerned over potential scale effects on long pointed fuselage shapes as a result of the XF8U-1 experiences in the Spin Tunnel, as discussed earlier. Thus, interest was growing over the possible use of much larger models than those used in spin tunnel tests in order

Radio-controlled drop models of the XF8U-1 and the X-1E being prepared for studies of incipient spin behavior using the helicopter drop-model technique in 1958. Note the large flow direction vanes used on the nose boom of the XF8U-1 model. Instrumentation for the drop models was rapidly advanced and miniaturized in subsequent models.

to eliminate or minimize undesirable scale effects. Finally, a major concern arose for some airplane designs over the launching technique used in the Spin Tunnel. Because the spin tunnel model was launched by hand in a flat attitude with forced rotation, it would quickly seek the developed spin modes—a valuable output—but the full-scale airplane might not easily enter the spin because of control limitations, poststall motions, or other factors.

To meet the need for a testing technique for studying the incipient spin at a suitable scale, Langley developed the approach of using free-flying, radio-controlled models. The technique consisted of launching an unpowered model into gliding flight from the helicopter, controlling the model from the ground with emphasis on evaluating the effects of control inputs on the incipient-spin motions, and retrieving the model with a recovery parachute when the flight was completed. Models used for the helicopter-drop tests were typically twice as large as the models used in the Spin Tunnel (1/9-scale, compared with the 1/20-scale).

One of the first configurations tested, in 1958, to establish the credibility of the drop-model program was a 6.3-foot-long, 90-pound model of the XF8U-1 configuration.[2] With spin tunnel results in hand, the choice of this design permitted correlation with the earlier tunnel and aircraft flight-test results. As has been discussed, wind tunnel testing of the XF8U-1 fuselage forebody shape had indicated that pro-spin yawing moments would be produced by the fuselage for values of Reynolds number below about 400,000, based on the average depth of the fuselage forebody. The Reynolds number for the drop-model tests ranged from 420,000 to 505,000, where the fuselage contribution became anti-spin, and the spins and recovery characteristics of the drop model were similar to the full-scale results. In particular, the drop model did not exhibit the flat spin mode predicted by a smaller spin tunnel model. The drop-model tests agreed with results of the aircraft flight tests. In addition to demonstrating the value of larger models from a Reynolds number perspective, these pioneering tests provided valuable information on the use of ailerons for prompt recovery of the XF8U-1 during poststall motions and the incipient spin.

NASA successfully conducted the XF8U-1 drop tests at the West Point, VA, airport, and the site quickly became active with investigations of other configurations. In 1959, the development of the X-15 research airplane had proceeded to include free-flight testing in the Full-Scale Tunnel, spin tunnel tests, and wind tunnel static tests over the range of Reynolds numbers corresponding to model flight tests. Results of the static tunnel testing had indicated that significant aerodynamic differences existed between all models (including a proposed drop model) and the full-scale aircraft for angles of attack greater than about 20 degrees. However, little was known about the relative effectiveness of the all-movable horizontal tails that were used for roll control on the X-15 for terminating out-of-control motions. Therefore, a drop-model study was undertaken to determine the behavior of the configuration during large-amplitude motions in the high-angle-of-attack region, where stalls, directional divergences, and spins were likely to be encountered.[3] The results of the drops indicated that the all-movable tail surfaces were effective for angles of attack below about 20 degrees, and although the model motions for much higher angles of attack did not necessarily represent those of the airplane, the application of roll control provided by the tail provided rapid recoveries for those conditions.

However, on flights in which the X-15 drop model was subjected to rapid pitching rates with simultaneous roll inputs, divergent conditions were encountered as a result of inertial coupling between lateral and longitudinal motions. Incipient spins followed the divergence, and the researchers demonstrated that roll control inputs in the direction of rotation could recover the model. Inertial coupling had been encountered in other aircraft development programs, and the X-15 design team appreciated the potential seriousness of the condition. Together with the results of the earlier wind tunnel free-flight model tests, the dynamic model tests at Langley provided a high level of confidence that conventional wing-mounted ailerons would not be required for the X-15 and that the all-moving tail concept was viable.

Success in applications of the drop-model technique for studies of spin entry led to the beginning of many military requests for evaluations of emerging fighter aircraft. In 1959, the Navy requested an evaluation of the McDonnell F4H-1 Phantom II airplane using the drop technique at West Point.[4] Earlier spin tunnel tests of the configuration indicated the possibility of two types of spins; one was steep and oscillatory, from which recoveries were satisfactory, and the other was fast and flat, from which recovery was difficult or impossible. As mentioned, the spin tunnel launching technique had led to questions regarding whether the airplane would exhibit a tendency toward the steeper spin or the more dangerous flat spin. The objective of the drop tests was to determine if it was likely, or even possible, for the F4H-1 to enter the flat spin.

In addition to following the general drop-model test procedure of launching an unpowered model from a helicopter in forward flight and controlling the model remotely from ground stations, researchers used an additional launching technique in an attempt to obtain a developed spin more readily and to obtain the flat spin. This technique consisted of prerotating the model on the helicopter launch rig before it was released in a flat attitude with the helicopter in a hovering condition. To reach even higher initial rotation rates than could be achieved on the launch rig, a detachable flat metal plate was attached to one wingtip. After the model appeared to be rotating sufficiently fast after release, the vane was jettisoned by the ground-based pilot, who, at the same time, moved the ailerons against the direction of rotation to help promote the spin. The model was then allowed to spin for several turns after which recovery controls were applied. In many aspects, this approach to testing replicated the spin tunnel launch technique but at a larger scale.

The launch helicopter with the drop model of the F4H-1 in an extended position on its launch rig prepares to climb to altitude at West Point in 1960. The primary objective of the tests was to establish the relative ease with which the model would enter a flat spin.

Results of the drop-model investigation for the F4H-1 are especially notable because it established the value of the testing technique to predict spin tendencies as verified by full-scale results. A total of 35 flights were made, with the model launched 15 times in the prerotated condition and 20 times in forward flight. During these flights, poststall gyrations were obtained on 21 occasions, steep spins were obtained on 10 flights, and only 4 flat spins were obtained. No recoveries were possible from the flat spins, but only one flat spin was obtained without prerotation. The conclusions of the tests stated that the aircraft was more susceptible to poststall gyrations than it was to spins; that the steeper, more oscillatory spin would be more readily obtainable; that recovery could be made by the recommended control technique; and that the likelihood of encountering a fast flat spin was relatively remote. Ultimately, these general characteristics of the airplane were replicated at full-scale test conditions during spin evaluations by the Navy and Air Force.

In 1960, the Navy once again requested drop-model tests, this time for the emerging North American A3J (later redesignated A-5 or RA-5) Vigilante attack aircraft.[5] The A3J configuration incorporated several unconventional features that might have affected its spin and recovery, including an all-movable horizontal tail for pitch control, a wing spoiler-deflector combination for roll control, and an all-movable vertical tail for yaw control. As was the case for the F4H-1, the spin tunnel test results for the A3J had indicated the possibility of a fast flat spin, from which recovery was difficult or impossible, and the issue of whether the airplane could easily enter the spin had initiated the Navy request. Once again, prerotating the model and releasing it from the helicopter at hovering conditions was used to promote spins. Only 9 spins were

A busy day of testing for the drop-model test unit at West Point in 1960. Models being prepared for tests include models of, front to rear, the Navy A3J attack fighter, a NASA Langley hypersonic reentry shape, and the Navy F4H-1 fighter.

obtained in 22 attempts, with 4 spins resulting from prerotated launches. However, satisfactory spin recoveries were not obtained from any of the spins. These results were immediately transferred to the airplane flight-test team and ultimately to the operational fleet, emphasizing the need to recognize the onset of a spin in its early stages and to terminate any turning tendency after the stall by rolling the airplane in the same direction as the turn.

Langley also conducted many spin, recovery, and incipient-spin studies of various versions of the Lockheed F-104 Starfighter during its development cycle. Several models with various configuration changes had been tested in the Spin Tunnel, and a one-seventh-scale model was used for drop-model testing.[6] Results of the spin tunnel tests and the drop-model tests indicated that it would be difficult to obtain a developed spin for the F-104. In the case of the helicopter-drop model, 11 attempts to spin the model by prerotating it and releasing it in a spinning attitude from the helicopter in a hovering condition resulted in only 2 spins. The model predictions agreed well with the full-scale airplane characteristics, the most significant factor being the difficulty of obtaining a developed spin.

Beginning in the early 1960s, a flurry of military aircraft development programs resulted in an unprecedented workload for the drop-model personnel.[7] Support was requested by the military services for the

General Dynamics F-111, Grumman F-14, McDonnell-Douglas F-15, Rockwell B-1A, and McDonnell-Douglas F/A-18 development programs. In addition, drop-model tests were conducted in support of the Grumman X-29 and DARPA-sponsored X-31 research aircraft programs, which were scheduled for high-angle-of-attack, full-scale flight tests at the Dryden flight facility. The specific objectives and test programs conducted with the drop models were different for each configuration. Spin tunnel testing had indicated that both the F-111 and F-14 would exhibit unrecoverable flat spins, and as the development of the full-scale aircraft proceeded, it became critical to determine the susceptibility of the configurations to enter the flat spin and to determine recommendations for recovery after loss of control and the incipient spin motions.

The F-111 drop tests in 1965 showed that the model would enter the flat spin relatively easily after misapplication of controls at poststall angles of attack. Once the model entered the flat spin, recovery was impossible, as had been experienced during the spin tunnel tests. Nearly every control combination was evaluated for recovery, including sweeping the wings forward and rearward during the recovery process. After over 50 successful tests of two 1/9-scale F-111 drop models, recommendations for recovery control procedures were forthcoming and provided to the industry-military team before the full-scale aircraft flight program. The spin susceptibility characteristics predicted by the drop models were verified in the flight program, which had benefited by the early indication that the configuration would readily enter a flat spin. During its service life, the F-111 was modified to include a stall-inhibitor system within its flight control system to help protect the pilot from inadvertent excursions into poststall conditions that might precipitate spins.

A one-ninth-scale model of the F-111 awaits launch from the helicopter over the Plum Tree Island test site in 1968. Two models were used in the program, which included follow-on NASA research on automatic spin prevention systems after direct support of the F-111 development efforts were completed.

As frequently happened in joint NASA-military programs, after the immediate aircraft development support activities had been completed, NASA retained the models (which had been fabricated with military funds) for general research. In 1969, a one-ninth-scale F-111 drop model was the subject of the first focused NASA research on automatic spin prevention.[8] The use of an automatic control system to sense and apply spin recovery controls was not new, but the concept would require special sensors and control system logic that would be used infrequently and would probably be maintenance intensive and failure prone. In the late 1960s, however, aircraft were on the drawing boards, with digital flight controls providing an unprecedented opportunity for sensing out-of-control situations and automatically applying recovery controls more rapidly than could a disoriented pilot. In addition, the special sensors that would be required to detect loss of control at high angles of attack were routine elements for conventional flight. Thus, special spin prevention systems would not require unique sensors that might be used infrequently.

However, considerable skepticism existed in the military community regarding the feasibility of automatic departure and spin prevention. The concept of "taking control away from the pilot" would require extensive demonstrations to the test and evaluation communities. In a special project, NASA designed control system logic algorithms and implemented an automatic spin prevention system for the F-111 drop model. Analytical studies, piloted simulator evaluations, and drop testing of the F-111 model were coordinated for an evaluation and demonstration of the effectiveness of automatic systems. Drop tests of the basic configuration without the automatic system showed that the model would enter a spin quite easily (for instance, with only longitudinal control). A fast flat spin was encountered, from which recovery could not be effected. With the automatic spin prevention system engaged, 19 attempts at spin entry were made during the drop tests, and no spins ever developed, even though the pilot maintained full pro-spin rudder and aileron deflections.

The activity to evaluate embryonic automatic spin prevention with the F-111 model was an unqualified success and quieted the skeptics, with the test results rapidly disseminated to the industry and military. Building on this fundamental study, designers of highly maneuverable aircraft have routinely incorporated special control system logic in digital flight control systems for enhanced carefree behavior at high angles of attack.

Early spin tunnel tests of the F-14 in 1970 revealed that the configuration might exhibit a fast flat unrecoverable spin mode and that the flat spin could not be terminated, even with an acceptable-sized parachute. This spin mode was deemed especially dangerous because the airplane might rotate rapidly (about 2 seconds per turn) about a vertical axis through its center of gravity, while descending vertically with the fuselage and a relatively horizontal attitude. Because of the high rate of rotation, the g-forces at the cockpit location would be high (approximately 6.5 longitudinal g's outward) and would probably incapacitate the pilot. After extended spin tunnel testing, it was found that a combination of anti-spin controls, a parachute with the maximum diameter acceptable, and foldout auxiliary canards on the fuselage forebody might provide for a marginally satisfactory recovery from the flat spin.

The issue of how easily the airplane might enter the unrecoverable spin once again became high priority for the Langley staff. Time was of the essence in addressing this issue for the F-14, and a two-pronged approach to drop-model tests was taken. It was decided to instrument two models that were already under fabrication so as to determine how easily would the airplane enter the flat spin and what control inputs were required to recover at various stages of the incipient spin. The first model was designed to quickly address the issues. It was equipped with a minimum of instrumentation, thereby avoiding unnecessary delays required to calibrate instruments, download data, and recalibrate data. The experienced drop-model team reverted to the earlier technique it had used for previous models at West Point, in which a miniature movie camera was carried onboard in one of the model's engine inlets and pointed forward so it could record the attitudes of vanes mounted to a nose boom for indications of angle of attack and angle of sideslip. Using this approach, the Navy could be provided answers in weeks rather than months. The second model was outfitted with sophisticated electronic instrumentation to provide quantitative information on angular rates, linear accelerations, control positions, and other flight variables. Testing of the second model could be conducted more leisurely to obtain quantitative data for analysis and correlation with analytical and simulator results.

The two-model approach to testing for the F-14 had been recommended by the Langley staff and approved by the Navy and Grumman as an expedient method to provide accurate information. Drop testing of the first F-14 model was soon underway after an intensive high-priority fabrication and outfitting process, providing an early indication of the spin resistance of the airplane. Initial flight tests involved applications of aggravated pro-spin control inputs and resulted in entry into the unrecoverable flat spin. Repetitive tests showed that the model could easily be flown into the spin; however, results indicated that the pro-spin inputs would have to be maintained for almost two turns before the spin became fully developed. Thus, corrective control inputs applied early in the motions were effective in promoting recovery back to conventional flight.

Using the simplified model technique, the Langley staff was productive, with minimal turnaround time required. For example, during a single test day, the team conducted 35 spin entry evaluations. A few months after the simplified model testing began, the more-instrumented second model entered testing, obtaining data that were invaluable for analysis and inputs to analytical studies and other analyses. Combining these data with static and dynamic wind tunnel aerodynamic data generated during conventional tunnel tests of F-14 models, piloted simulator studies of the high-angle-of-attack characteristics of the F-14 were conducted in the Langley Differential Maneuvering Simulator in 1972 with a NASA-designed automatic departure and spin prevention concept.[9] The basic control concept developed in the NASA simulator studies underwent evaluation in full-scale F-14 flight tests in a joint NASA–Navy–Grumman program at Dryden. After a lengthy gestation, a derivative of the concept was applied to the F-14 fleet in a new advanced digital flight control system in the 1990s.

The F-14 aircraft was used for the joint NASA–Grumman–Navy assessment of automatic control concepts for enhanced high-angle-of-attack behavior. NASA's use of free-flight models in the Full-Scale Tunnel, the Spin Tunnel, and the drop-model operation provided substantive data for the planning of the program.

Almost simultaneous with the F-14 activity, an Air Force request for drop-model evaluations of the F-15 was received. The F-15 had been designed with special consideration of spin resistance, and the configuration included mechanical aileron-rudder control interconnects to enhance maneuverability at high angles of attack while maintaining resistance to inadvertent spins. Analytical studies and NASA free-flight wind tunnel tests in the Langley Full-Scale Tunnel had indicated that the design objective had been met and that the aircraft would exhibit spin resistance. The objective of the drop-model program, therefore, had an unconventional theme, in that the researchers attempted to discover control manipulations and aggravated applications that would overpower the inherent spin resistance of the F-15 design and promote spins. Initial results of drop-model tests at Plum Tree in 1971 validated the effectiveness of the airframe and control system of the F-15, which was judged to be one of the more spin resistant aircraft tested in the long series of Langley evaluations. Initially frustrated by an inability to spin their F-15 drop model, the Langley researchers worked out a control input sequence that resulted in spins for a constrained set of control conditions. A particularly productive technique involved movement of the control stick while full roll control was applied to promote the spin. When the technique was applied in the drop tests, spins could be produced at will. This control technique was adopted in the full-scale F-15 flight program for promoting spins.

During the early 1970s, the NASA Dryden Flight Research Center had begun efforts to develop a sophisticated remotely piloted research vehicle (RPRV) testing technique wherein relatively large models were launched from a B-52 mother ship and controlled from a realistic ground-based cockpit by a research test pilot. After the Langley spin tunnel and drop-model tests were underway, the Dryden staff conducted a large-scale unpowered model drop test in 1973 to evaluate the departure and spin characteristics of the F-15, and to demonstrate the application of the drop technique at a larger scale.[10] The drop model used for the tests at Langley was a 13-percent-scale model, whereas the model used for the Dryden study was a 38-percent-scale model.

Initial results of the Dryden flights for the basic F-15 verified the departure and spin resistant characteristics shown by the smaller model at Langley. In the Dryden study, the model flight tests were augmented by piloted simulator studies based in part on aerodynamic data extracted from the flights. Using the simulator, Dryden researchers verified Langley's techniques to defeat the spin resistance features of the flight control system. When results between the 13-percent and 38-percent models were correlated with the airplane results, it was found that the agreement was good, including control strategies to overcome the inherent spin resistance of the configuration.

A drop-model study at Langley in 1974 in response to an Air Force request to support development of the B-1A bomber was short-lived, consisting of only a few flights. With its wings swept to moderate and aft-sweep positions, the configuration exhibited a severe longitudinal instability in the form of an abrupt pitch-up at high angles of attack. For this particular aircraft configuration, the free-flight test in the Full-Scale Tunnel was sufficient to define the critical problems of the design, and the drop-model program was relatively unproductive. A stall inhibitor system was incorporated in the follow-on B-1B flight control system to limit the achievable angle of attack to values within the flight envelope.

The 0.38-scale F-15 RPRV lands on the lakebed at NASA Dryden after a successful flight.

A Navy request in 1978 for drop-model testing to define the spin resistance characteristics of the F/A-18A resulted in a significant upgrade to the Langley testing technique. When the F/A-18A was developed as one of the first digital fly-by-wire configurations, it became mandatory to represent its complex high-angle-of-attack control logic and control surface scheduling. As an investment in the F/A-18A development and its own future interests, the Navy funded an upgrade for the Langley drop-model test technique based on ground-based digital computers for simulation of advanced flight control laws. The technical results of the program verified the high degree of spin resistance of the F/A-18A and contributed to flight-test guidance and knowledge.

NASA support of the F/A-18A development program included a major upgrade of the ground-based flight control system. A digital computer was introduced for simulation of the critical components of the control system for the full-scale airplane.

In addition to the support testing of specific DOD aircraft, NASA conducted drop-model tests of the X-29 and X-31 research aircraft.[11] In 1987, Langley once again upgraded its drop-model test technique for testing of the X-29 configuration. Langley had never flown an unstable drop model before the X-29 activity. Nearly every element of the testing technique, including model control actuators, transmitters and receivers, data importers and decoders, and operational displays had to be updated to fly the sophisticated, inherently unstable aircraft. The challenges to conduct the activity were the most significant ever posed to the Langley drop-model team. The results of the X-29 drop program demonstrated that advanced control systems would play a large role in poststall motions and spin resistance. In the case of the X-29, without artificial stabiliza-

tion in roll, the drop model demonstrated large-amplitude rolling oscillations that became unstable at angles of attack above 30 degrees, followed by a departure and poststall gyrations. However, when elements of the full-scale control system were appropriately represented, the model was resistant to intentional spins and was maneuverable at high angles of attack. In addition to providing support and guidance for the full-scale flight-testing, Langley conducted fundamental research with the drop model, including the extraction of aerodynamic parameters during the complex rolling motions exhibited by the unaugmented airplane.

The X-31 drop model is inspected in 1991. The drop program ran simultaneously with the full-scale flight program being conducted at Dryden, providing real-time support for problems that occurred in flight.

Some of the drop models used in support of the Nation's military programs are shown in this photograph of 1994 with lead Langley researcher Mark Croom.

The 0.27-scale drop model used in support of the X-31 program at Langley was larger than any drop model previously tested, weighing over 540 pounds. This particular drop model was key during full-scale flight-test evaluations of the X-31 aircraft at Dryden in the early 1990s. Two problems that surfaced during the flight tests had resulted in delays in the flight program while solutions were found. The first problem involved a lack of sufficient nose-down aircraft response to control inputs at extreme angles of attack. In support of the project, Langley conducted wind tunnel testing in the Full-Scale Tunnel to define a pair of fuselage afterbody strakes that promoted nose-down recovery. The second problem involved uncontrollable large asymmetric yawing moments by the aircraft at high angles of attack. Once again, Langley conducted tunnel tests and provided a solution involving nose strakes on the forebody. Both the forebody and afterbody strakes were evaluated and demonstrated using the X-31 drop model in almost real-time fashion and adapted for the full-scale aircraft, which demonstrated unprecedented maneuverability at extreme angles of attack. The X-31 drop-model program was arguably one of the more valuable tools used to ensure the success of this revolutionary aircraft.

Most of the drop-model activities in this section were conducted at Langley's satellite test site at nearby Plum Tree Island. After the X-31 project, NASA closed its Plum Tree site in 1995, restored the area to its original wildlife refuge environment, and conducted extensive studies of alternate locations for drop-model studies. When the Navy began development of its F/A-18E/F Super Hornet, Langley tested a sophisticated drop model at the NASA Wallops Flight Facility to provide risk reduction for the high-angle-of-attack part of the full-scale flight-test program. The tests, which used a large 0.22-scale model required additional changes to the testing technique hardware and procedures.[12]

The F/A-18E model used in the test at Wallops weighed between 880 to 1,100 pounds for various test conditions and was dropped from a special helicopter at altitudes of about 15,000 feet (compared to about 3,000 feet for the earlier Plum Tree test programs). During flights, the model was tracked by a manually operated ground-based tracker equipped with video cameras, a radar-ranging system, and telemetry antennas. A camera was mounted in the cockpit to provide an onboard view of the flight. Data generated by the drop-model program covered more areas than past programs had. For example, in recent years, concern had arisen over the ability of highly maneuverable fighters to generate crisp nose-down control for recovery from high angles of attack. In the drop program, the model was stabilized near maximum lift, and a rapid nose-down control was applied, resulting in good correlation with response characteristics exhibited by the airplane.

Spin resistance and recovery studies included assessing the effectiveness of the sophisticated F/A-18E/F control system, which includes extensive elements for pilot-directed spin recovery inputs. The model (and the airplane) exhibited a reluctance to spin after conventional spin entry maneuvers. Instead, large amplitude oscillations began and produced self-recovery from the spin attempt. Extensive evaluations of the impact of entry rates on spin resistance were conducted, and unusual poststall gyrations known as the "cartwheel" and "falling leaf" were exhibited with good correlation with spin tunnel tests and full-scale flight results. Quantitative correlation of all drop-model results with piloted simulator results and data from full-scale flight tests was an intimate part of this program. The program also investigated both upright and inverted departure and spin characteristics with and without asymmetric mass loadings. Without doubt, the F/A-18E drop-model program was the most sophisticated and extensive technical study ever undertaken using a drop-model technique at Langley.

General-Aviation Configurations

In the early 1970s, fatal stall/spin accidents within the general-aviation community had increased at an alarming rate, and it was widely recognized that spin research for this type of aircraft had not kept up with the pace of new airplane designs for almost 20 years. In recognition of this void in technology, Langley started a general-aviation spin research program in 1972, with the primary objectives of determining how various parameters affect spin and recovery characteristics, providing data on parachute size required for emergency spin recovery for typical light airplanes during spin tests, developing a radio-control powered-model testing technique that a manufacturer could utilize in aircraft development, and validating model test results by correlation with full-scale airplane results.[13] The configurations selected for the program included low and high wing designs considered representative of modern light general-aviation airplanes. Although based on specific airplane configurations, the designs to be studied were intentionally modified from the basic airplane with significant geometric changes.

A key element in the general-aviation spin program was the use of powered radio-controlled models to study spin resistance, spin entry, and spin recovery during the incipient phase of the spin. Equally important was a focus on developing a reliable, low-cost testing technique that the industry could use for predictions of characteristics in early design stages. The dynamically scaled models, which were about 1/5-scale (wingspan of about 4–5 feet), were powered and flown with hobby equipment. Although resembling conventional radio-control models flown by hobbyists, these scaled models were relatively heavy (about 15–20 pounds) compared with conventional hobby models (about 6–8 pounds).

The radio-controlled model activities in the Langley program consisted of three distinct phases. Initially, model testing and analysis were directed at producing timely data for correlation with spin tunnel and full-scale flight results to establish the accuracy of the model results in predicting spin and recovery characteristics, and to gain experience with the testing technique. The second phase involved assessments of the effectiveness of wing leading-edge modifications to enhance the spin resistance of several general-aviation configurations. The focus of this research was a concept consisting of a drooped leading edge on the outboard wing panel with a sharp discontinuity at the inboard edge of the droop. The third phase involved cooperative studies of specific general-aviation designs with manufactures and designers. In this segment of the program, studies centered on industry's assessment of the radio-controlled model technique. Model flight-testing for the cooperative programs was conducted near Langley at the West Point, VA, airport, the Langley Plum Tree test site, and the locations of manufacturers.

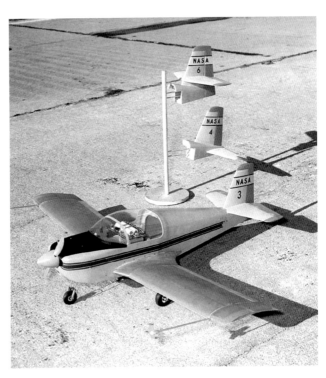

The NASA general-aviation spin research program of the 1970s included extensive correlation studies of results obtained with spin tunnel tests, radio-controlled model tests, and airplane flight tests. Many key parameters were studied, including the effects of tail configuration as shown for the low wing research model.

Direct correlation of results for radio-controlled model tests and full-scale airplane results for a low wing NASA configuration was good, especially with regard to susceptibility of the design to enter a fast flat spin with poor or no recovery.[14] In addition, the effects of various control input strategies agreed well. For example, with normal pro-spin controls and any use of ailerons, the radio-controlled model and the airplane were both reluctant to enter the flat spin mode that had been predicted by spin tunnel tests; they only exhibited steeper spins, from which recovery could be effected. Subsequently, the test pilot and flight-test engineers of the full-scale airplane developed a unique control scheme during flight tests that would aggravate the steeper spin and propel the airplane into the unrecoverable flat spin. When a similar control technique was used on the radio-controlled model, it also would enter the flat spin, requiring the use of the emergency parachute for recovery.

Arguably, the most impressive results of the radio-controlled model program for the low wing configuration related to the ability of the model to demonstrate effects of the discontinuous leading-edge droop

This T-tail low wing NASA research aircraft displays the discontinuous wing leading-edge modification that significantly increased the spin resistance of configurations in NASA's stall/spin program. Radio-controlled model results for another low wing configuration agreed well with full-scale flight results.

concept that had been developed by Langley for improved spin resistance.[15] Several wing leading-edge droop configurations had been derived in wind tunnel tests with the objective to delay wing autorotation and spin entry to high angles of attack. Tests with the radio-controlled model when modified with a full-span droop indicated better stall characteristics than did the basic configuration, but the resistance of the model to enter the unrecoverable flat spin was degraded. The flat spin could be obtained on nearly every flight if pro-spin controls were maintained beyond about three turns after stall.

In contrast to this result, when the discontinuous droop was applied to the outer wing, the model would enter a steep spin from which recovery could be attained by simply neutralizing controls. When the discontinuity on the inboard edge of the droop was faired over, the model reverted to the same characteristics that had been displayed with the full-span droop and could easily be flown into the flat spin. Correlation between the radio-controlled model and aircraft results in this project were outstanding. The agreement was particularly noteworthy in view of the large difference between the model and full-scale Reynolds numbers. All of the important stall/spin characteristics displayed by the low wing radio-controlled model with the full-span droop configuration and the outboard droop configuration (with and without the fairing on the discontinuous juncture) were nearly identical to those exhibited by the full-scale aircraft, including stall characteristics, spin modes, spin resistance, and recovery characteristics.

While conducting the technical objectives of the radio-controlled model program, researchers directed an effort at developing test techniques that might be used by industry for relatively low-cost testing. Innovative, low-cost instrumentation techniques were developed that used inexpensive onboard sensors to measure control positions, angle of attack, airspeed, angular rates, and other variables. Data output from the sensors was transmitted to a low-cost, ground-based data acquisition station by modifying a seven-channel radio-control model transmitter. The ground station consisted of separate receivers for monitoring angle of attack, angle of sideslip, and control commands. The receivers operated servos to drive potentiometers, whose signals were recorded on an oscillograph recorder. Tracking equipment and cameras were also developed.[16] Other facets of the test technique development included the design and operational deployment of emergency spin recovery parachutes for the models.

One particularly innovative testing technique demonstrated by NASA in the radio-controlled model flight programs was miniature auxiliary rockets mounted on the wingtips of models to artificially promote flat spins. This approach was particularly useful in determining the potential existence of dangerous flat spins, which were difficult to enter from conventional flight. In this application, one of the rockets was remotely ignited during a spin entry by the pilot, resulting in high spin rates and a transition to high angles of attack and flat spin attitudes. After the "spin-up" maneuver was complete, the rocket thrust subsided, and the model either remained in a stable flat spin or pitched down to a steeper spin mode.

General-aviation manufacturers maintained a close liaison with Langley researchers during the NASA stall/spin program, absorbing data produced by the coordinated testing of models and full-scale aircraft. The radio-controlled testing technique was of great interest, and after frequent interactions with Langley's test team, industry conducted its own evaluations of radio-controlled models for spin testing. In the mid-1970s, Beech Aircraft conducted radio-controlled testing of its T-34 trainer aircraft, the Model 77 Skipper trainer, and the twin-engine Model 76 Duchess.[17] Piper Aircraft also conducted radio-controlled model testing to explore the spin entry, developed spin, and recovery techniques of a light twin-engine configuration.[18] Later, in the 1980s, a joint program was conducted with the DeVore Aviation Corporation to evaluate the spin resistance of a model of a new high wing trainer that incorporated the NASA-developed leading-edge droop concept.[19] As a result of these cooperative ventures, industry obtained valuable experience in model construction techniques,

The Beech Aircraft Corporation conducted spin tests with this radio-controlled model of the YT-34 military trainer in 1976. Note the wingtip booms with flow vanes, the tail-mounted spin recovery parachute, and the wingtip rocket pods used for pro-spin inputs.

spin recovery parachute system technology, methods of measuring moments of inertia and scaling engine thrust, the cost and time required to conduct such programs, and correlation with full-scale flight-test results.

Interestingly, the results obtained by Langley in the radio-controlled model studies of discontinuous wing leading-edge droop have been reflected in recent model-airplane kits offered by manufacturers of conventional radio-controlled models. Some kit designers have included removable wing discontinuous droops in model kits as a safety enhancement to be used during the training of radio-control model pilots. The wing droops are provided as optional additions to the basic model wing configuration for enhanced stall and spin resistant behavior during training, after which they can be removed to permit full aerobatic maneuvers, including spin and recovery. The modifications have been referred to as "NACA droops," when in fact, they were developed in the NASA studies.

Langley researchers pose with a radio-controlled model used in a cooperative program on spin resistance with the DeVore Aviation Corporation. The model was equipped with discontinuous outboard droops and proved spin resistant. Note the low-cost data acquisition unit.

Jet Transport Upset Recovery

Detailed analyses of causal factors in fatal accidents of modern large commercial transport aircraft indicate that loss of control has been one of the leading contributors, usually involving losses of aircraft and lives. Many of the accidents have involved large excursions to poststall angles of attack or steep pitch and bank attitudes from which recovery may be difficult. As part of its Aviation Safety program, NASA initiated a project in 1999 to improve the understanding and technologies required to minimize such accidents. Because large attitude upsets to poststall conditions involve complex aerodynamic phenomena similar to those experienced in stall/spin motions and are difficult to mathematically model, dynamic models are a main element in the research. The NASA Airborne Subscale Transport Aircraft Research (AirSTAR) project was formulated to develop the test techniques and a flight facility for research and validation of technologies.[20] The objectives include defining static and dynamic aerodynamic data for transport aircraft configurations at high angles of attack and sideslip in wind tunnel tests, extracting aerodynamic data from flight motions, developing flight simulator databases for studies of the problem, and providing technologies for adaptive control applications that quickly recognize an out-of-control condition and enable recoveries from the situation.

The AirSTAR activity includes a subscale flight research vehicle, an evaluation pilot, and the safety pilot. To date, the research models have ranged from small propeller-driven configurations to relatively large 5.5-percent-scale models of generic transport configurations powered by turbine engines. The testing technique consists of a powered takeoff of the model from a runway while under control of the safety pilot using radio-control equipment. Once airborne at the test condition, control is passed to the research pilot seated in a mobile operations station. The research pilot uses a synthetic vision display and other graphic inputs to set up the test condition, and a ground-based flight control computer in the operation station provides inputs for control laws and interface with research pilot. All flights are conducted within view of the safety pilot, who can override the research pilot inputs in an emergency. At the end of the research task, the research pilot hands off control of the model to the safety pilot, who performs the landing.

Progress in the AirSTAR project has emphasized the development of the hardware, software, and skills required for the focused objectives of the research tasks. The latest model flight tests have been conducted with a commercial off-the-shelf transport model, which has been used in preparation for flights of a more expensive dynamically scaled model of a current jet transport configuration. An extensive aerodynamic database has already been gathered for the scaled model during tests in the Langley 14- by 22-Foot Subsonic Tunnel and is awaiting correlation with flight-derived data. The scope of current activities includes real-time system identification, advanced controls designed for recovery from upset conditions, and advanced methods for air data measurements. Flight-testing of the dynamically scaled model will incorporate lessons learned and risk mitigation strategies from the precursor evaluations of the commercially available transport model.

In preparation for model flight tests in the AirSTAR project, Langley has conducted conventional and dynamic force tests in the 14- by 22-Foot Tunnel to measure the aerodynamic characteristics of the configuration.

Tumbling

Since the earliest days of flight, it has been recognized that certain configurations might be susceptible to a phenomenon known as tumbling. Tumbling is characterized as an autorotative pitching motion in which the aircraft continually rotates 360 degrees in pitch while descending along an inclined flight path. For many configurations, the tumbling motion cannot be terminated with conventional aerodynamic controls, and the phenomenon must be avoided during excursions to high angles of attack. Flying wing, tailless, and statically unstable aircraft designs have been particularly susceptible to tumbling. The phenomenon has been suspected as the cause of the crash of the Northrop YB-49 in the late 1940s. The NACA conducted extensive studies of tumbling in the Spin Tunnel at Langley during World War II. The testing procedure consisted of releasing a model in the vertical airstream of the tunnel without rotation in a nose-up attitude simulating a "whip" stall to zero airspeed. Tests of a model of the flying wing Northrop XB-35 in the tunnel highlighted the unrecoverable nature of the tumbling motions, which could only be avoided by moving the center of gravity of the model forward.[21] Other configurations found susceptible to tumbling during spin tunnel tests included the tailless forward-swept wing Corneilius XFG-1 towed fuel glider, the tailless Douglas XF4D, and the Northrop X-4 research airplane.[22]

Interest in tumbling was renewed when the concept of relaxed longitudinal stability was introduced for radical configurations such as the Grumman X-29. The X-29 was designed to have a high level of inherent aerodynamic instability about the pitch axis. In conventional flight, a multichannel stability augmentation system deflected the longitudinal control surfaces to maintain a satisfactory level of stability. As part of its general investigation of spin and recovery characteristics of the X-29, Langley evaluated the tumble suscepti- bility of the configuration and the effectiveness of controls for recovery.[23]

A spin model of the X-29A is mount- ed on a free-to-pitch test appara- tus during studies of the tumbling resistance of the configuration in the Spin Tunnel. The test setup permit- ted 360-degree pitching motions without restraint.

Langley's research with radio-controlled parawing vehicle models demon- strated that that class of vehicle might also be susceptible to tumbling.[24] During flight tests designed to evaluate the general controllability and stalling characteristics of a parawing utility configuration, the model was found to have satisfactory stall behavior with no roll-off tendencies and a high degree of spin resistance, even with pro-spin controls applied by the pilot. However, after abrupt stalls ("whip stalls"), the model exhibited a violent longitudinal instability in which a strong initial nose-down motion was followed by an end-over-end tumbling motion. After the model was tumbling, the controls were ineffective in terminating the motion, and in one instance, the model completed approximately 20 to 25 tumbles before it crashed. A close examination of the model during the tumbling motions revealed that it pitched nose down so rapidly that the billowing of the parawing fabric reversed direction and caused the rate of pitch to increase. The Langley researchers determined that an early application of recovery control applied before the model reached a vertical nose-down attitude could effect recovery and avoid tumbling. This type of in-depth free-flight model study of the dynamic stabil- ity and control of parawing vehicles during large-amplitude maneuvers, coupled with wind tunnel force tests of the same model provided fundamental information and valuable guidance to designers and pilots. Even today, tumbling remains a significant concern for microlight recreational configurations.[25]

In view of emerging interest in flying wings and relaxed stability for civil and military aircraft and the lack of guidelines for avoiding tumbling, Langley engaged in a generic investigation based on experimental results obtained with tailless models in the Spin Tunnel.[26] The objectives of the program were to identify geometric and mass parameters that cause susceptibility to tumbling and to analyze some of the factors that cause steady-state tumbling motions. Twelve flying-wing models were tested in the Full-Scale Tunnel and the Spin Tunnel in a series of coordinated investigations, including in-depth measurements of static and dynamic aerodynamic parameters, free-tumble tests, and constrained free-to-pitch tests. The experimental study was augmented with analyses of aerodynamic characteristics, and the results of the integrated pro- gram were summarized in terms of design variables such as wing aspect ratio, longitudinal stability, and mass distribution.

Other Studies

Over the years, free-flight models have been applied in other studies of poststall flight dynamics of inter- est to the aviation community. Two examples of such applications involved supplying critical data to acci- dent investigation boards. The first example was conducted after the fatal accident of a British BAC 111 jet transport in a fatal accident August 6, 1966, near Omaha, NE.[27] The aircraft was seen to burst into flames at an altitude of about 5,000 feet and crash in a flat attitude without significant forward motion. Analysis indicated that the breakup of the aircraft had been caused by an abrupt, powerful asymmetrical gust that exceeded the design requirements of the T-tail structure, and that the aircraft had pitched over to a large angle of attack, causing the outer wing panel to break off. Langley spin tunnel engineers built a crude model of the damaged airplane and proceeded to launch it from the 250-foot-high top of the Langley Lunar Landing Facility. After being catapulted, the model tumbled violently, then settled into a slow flat spin and impacted in a flat attitude similar to the full-scale aircraft. These experimental results helped the accident board arrive at its conclusions.

Another example of the use of dynamically scaled models to provide a quick response to accident investigations was spin tunnel testing of the North American XB-70 configuration. At the request of the Air Force in 1965, tests had been conducted in the Spin Tunnel and at a catapult launch facility in a large airship hangar at the Weeksville Naval Air Facility at Elizabeth City, NC, to determine the potential spin entry and recovery characteristics of a 1/60-scale model of the XB-70.[28] After the accident of XB-70 Ship 002 after an in-flight collision at Edwards Air Force Base during a photo mission June 6, 1966, additional tests were conducted by Langley with the 3-foot-long model of the XB-70 in the Spin Tunnel to assess its spin characteristics in the damaged configuration after the collision.[29] Of particular interest was the question of whether g-loads experienced at the cockpit location during a flat spin that followed the collision could have prevented ejection from the airplane. For these tests, the XB-70 model was first tested in its basic undamaged configuration. It was then reconfigured to represent the damaged full-scale XB-70 as closely as possible, including removal of both vertical tails and the outer part of the left wing panel. Results of the tests showed that the basic configuration would exhibit a moderately flat spin and that the damaged configuration would spin even faster and flatter. The data from the Spin Tunnel were transmitted to the investigation board for use in its deliberations.

Summary

The application of powered and unpowered free-flight models in studies of poststall motions such as spin entry and tumbling has provided insight into dynamic flight behavior that could not have been easily analyzed with other wind tunnel techniques. Properly integrated with piloted simulator studies, the techniques have been used to provide realistic predictions of the relative resistance of configurations to enter potentially dangerous modes of motion. They have also been used to demonstrate the effectiveness of special control systems to prevent undesirable motions and provide "carefree" maneuvers.

The development of relatively low-cost testing techniques that can be used by a wide segment of the aviation community has been particularly valuable for designers of light personal-owner aircraft, and the experience gained with remotely piloted techniques for more sophisticated research models has pointed the way toward more quantitative information from model flight tests.

CHAPTER 8:

ASSOCIATED TEST TECHNIQUES

Aerodynamic Data

Arguably, the most formidable challenge facing the stability and control analyst is an accurate prediction of the aerodynamic properties of the aircraft under consideration. Aircraft weight-and-balance characteristics such as moments of inertia and center-of-gravity location are straightforward to quantify. However, aerodynamic data are typically complex and dependent on configuration details, dynamic motions, and effects of Reynolds number and Mach number. Regardless of their complexity, aerodynamic characteristics are the key factor in the analysis of stability and control.

One of the more significant advantages of free-flight models is the possibility of measuring the aerodynamic characteristics of the model at the same value of Reynolds number used in the model flight tests. Potential scale effects are at least eliminated for the model study, and aerodynamic trends can be interpreted with confidence at model scale. An understanding of model trends is extremely helpful in interpreting data obtained at higher Reynolds numbers. Obtaining aerodynamic data before conducting free-flight tests provides early inputs for identification of potential problems in stability and control characteristics that might be expected during the model flight tests. Correlation of the model's aerodynamic data with results obtained from other wind tunnels, if available—especially for higher values of Reynolds number—ensures that the data magnitudes and trends are reasonable before flight-testing begins. Methods used to extract aerodynamic data from model flight tests via systems identification techniques can provide guidance when similar methods are used at other flight conditions not directly addressed in the model flight-test program. After the aerodynamic data are analyzed and correlated with observations during free-flight model tests, they are typically used as inputs in more sophisticated analysis techniques such as piloted simulators. Used in this manner, the free-flight aerodynamic data provide credible inputs and realism for pilot evaluations and training.

In addition to conventional static wind tunnel force and moment tests, NASA has conceived and developed unique dynamic wind tunnel test procedures to measure aerodynamic parameters caused by dynamic motions of aircraft. NASA has also applied other dynamic testing, in which the model is permitted a single degree of freedom, to study certain types of aircraft instabilities that occur for high angles of attack or separated flow conditions.

A review of NASA activities for the development and application of these wind tunnel aerodynamic testing techniques for free-flight models follows; however, the reader is referred to the excellent summary by Owens, et al.,[1] for an in-depth review of the methods.

Static Tests

Whenever possible, a wind tunnel free-flight program begins with conventional static aerodynamic force and moment measurements before the model is outfitted for flight tests. For static wind tunnel testing, an electrical strain-gauge balance is mounted within the fuselage of the free-flight model to provide six-component measurements of forces and moments. Longitudinal measurements include normal force, axial force, and pitching moment, while lateral-directional measurements include side force, rolling moment, and yawing moment. Control surfaces and configuration variables such as wing leading- or trailing-edge flaps are manually set at prescribed deflections, and the angle of attack and angle of sideslip during the conventional static force and moment tests are varied over a large range, including poststall conditions. The test engineer examines the measurements for obvious problems and trends, such as aircraft instabilities or sudden deterioration in control effectiveness. Candidate solutions for the problems are identified, and a free-flight program is planned with these issues in mind. Certain free-flight configurations, such as V/STOL aircraft designs, are subject to large power-induced effects, and power-on force tests are mandatory before flight. The determination of power effects complicates the test setup and scope of the aerodynamic test program.

The large, heavy outdoor drop models used by NASA for spin entry testing are usually not subjected to preflight tunnel testing because of unacceptable delays in the flight schedule; however, a few models have undergone testing to determine control hinge moments, calibrate air data systems, or provide information on other operational issues.

In many instances, researchers have found that an abrupt onset of instabilities experienced in the free-flight tests was not predicted within the necessarily limited number of test conditions during the preflight static testing. On those occasions, additional wind tunnel entries were required to expand the aerodynamic database with additional test points at finer increments of angle of attack and sideslip. These experiences were particularly prevalent for investigations of high-angle-of-attack behavior near stall conditions. In rare instances, the flexible flight cable used in the model flight tests was found to create an unacceptable aerodynamic interference effect on stability of certain configurations. For example, in one case, the presence of the flight cable above the fuselage of an advanced fighter configuration severely degraded lateral stability because of interference effects of the cable's pressure field with critical vortical flow on the upper fuselage. Additional static testing was conducted to find an acceptable location for repositioning the cable. Typically, the solution consisted of bringing the flight cable high-pressure air tubes and control cables into the model from its underside while maintaining the upper surface location of the steel safety cable. In some cases, the foregoing static wind tunnel aerodynamic testing was integrated with flight tests over the duration of the study, with the static test conducted before, during, and after flight tests.

Forced-Oscillation Tests

Analysis and prediction of aircraft dynamic stability and control requires knowledge of both static and dynamic aerodynamic phenomena. Dynamic aerodynamic phenomena include an understanding of aerodynamic effects caused by motions such as angular rates in pitch, roll, and yaw. For most configurations, the most important dynamic parameters are the damping derivatives, which indicate the magnitude of aerodynamic resistance exhibited by the configuration during rotary motions. These parameters are of particular interest to the stability-and-control analyst, because aerodynamic damping can be either stable, retarding moments or unstable moments that augment the motion.

The NACA and NASA developed a test technique known as forced-oscillation testing, during which the model and its internal strain-gauge balance are mounted to special apparatuses that force the model to move in angular motions of specified frequency and amplitude.[2] The electrical output of the balance during the oscillatory motions is routed to a special analyzer, which interrogates the return signal in terms of forces and moments in phase with the angular displacement of the model, and also in terms of forces and moments in phase with the angular rates of the model. These measurements are then converted to engineering values of aerodynamic forces and moments about each aircraft axis. For example, during a forced-oscillation test in roll, parameters are measured for the rolling moment caused by rolling (roll damping), the yawing moment caused by rolling, and the side force caused by rolling. The reduced data are then analyzed for trends in dynamic aerodynamic behavior, with special attention given to sudden changes in characteristics. Most aircraft configurations exhibit stable aerodynamic damping in roll during rolling motions at low angles of attack and attached flow conditions. As the angle of attack is increased to conditions near stall, massive flow separation can occur on the wing, resulting in propelling, rather than damping, variations in damping in roll. Such trends may lead to

A typical test setup for forced-oscillation tests in pitch is shown for a free-flight model of the F-15 in the Langley 14- by 22-Foot Tunnel. An electric motor at the base of the support column drives a flywheel-pushrod-crank assembly, which forces the model and its yoke-mounted internal strain-gauge balance to oscillate in pitch at a specified amplitude and frequency.

lightly damped or unstable roll oscillations at high angles of attack, requiring the use of artificial stabilization typically provided by roll-rate sensors and control system logic. An early detection of problems requiring control system architecture changes is valuable in early design development of advanced aircraft.

Forced-oscillation testing has been a common element in wind tunnel free-flight testing, and rigs have been developed and used in tests of free-flight models in Langley's 12-Foot Low-Speed Tunnel, Full Scale Tunnel, and 14- by 22-Foot Tunnel. Nearly every military free-flight model at Langley has been tested using the technique, and the data derived were typically incorporated in flight-test planning for the full-scale air-craft, including the aerodynamic defi-nition of the configuration used in piloted simulators. The dynamic data have also been used for the calibra-tion of computational methods for estimating dynamic derivatives. In addition to the low-speed applica-tions typically involved in flight dynamics research on free-flight models, Langley has developed high-speed forced-oscillation test rigs to permit measurements of dynamic stability derivatives at tran-sonic and supersonic speeds with conventional metal models in the Langley Transonic Dynamics Tun-nel, the Langley National Transonic Facility, and the Langley Unitary Supersonic Wind Tunnel.

For forced-oscillation tests in roll, the assembly at the top of the vertical col-umn is replaced with a mechanism that permits the pushrod motion to be converted to rolling oscillations. The photograph shows a free-flight model of the F/A-18E undergoing roll oscillation testing in the 14- by 22-Foot Tunnel. For yawing forced-oscillation tests, the sting is reoriented to enter the model through the top or bottom.

Free-to-Roll Tests

As experience with free-flight models and corresponding full-scale aircraft was accumulated, researchers experimented with other approaches to identify potential roll damping issues, particularly for high-angle-of-attack conditions. Many slender, swept wing configurations that exhibited large-amplitude, lightly damped roll oscillations in flight seemed to exhibit such motions about the longitudinal aircraft axis—that is, almost pure rolling motions. In an attempt to replicate the rolling motions, dynamic models were sting-mounted to a minimum-friction bearing assembly, which provided the model 360-degree freedom in roll at a specified angle of attack. The technique, known as "free to roll," was calibrated by testing models that had exhibited roll oscillations in flight tests in the wind tunnel. One of the first models tested with a free-to-roll rig was the F-4 fighter that, as discussed, exhibited large-amplitude wing-rocking motions at high angles of attack.

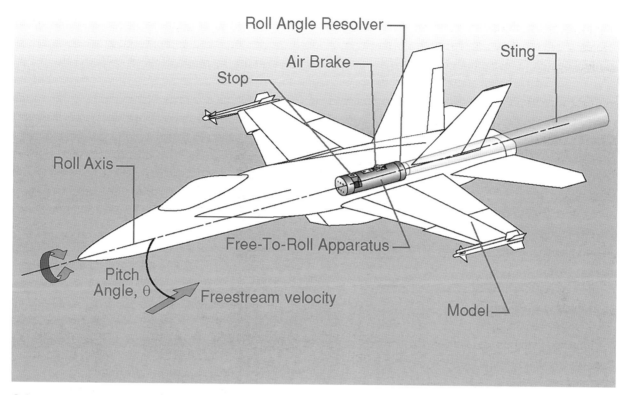

Schematic of a typical test setup for free-to-roll tests. The model is mounted to a special bearing-equipped apparatus that permits 360 degrees of freedom in roll. A brake is provided to lock the roll attitude if desired.

Results of the free-to-roll tests showed good correlation with the F-4 free-flight tests. Comparison of free-to-roll results with measurements of roll damping obtained from forced-oscillation roll tests for the F-4 model indicated that the free-to-roll trends followed the roll-damping measurements, thereby providing guidance for analyses of aerodynamic phenomena. The most significant advantage of the free-to-roll testing technique was that it offered the potential for a test setup in which the roll degree of freedom could be constrained, thereby permitting the test apparatus to be used for conventional testing. In this manner, free-to-roll characteristics might be obtained during a single tunnel entry that combined static and dynamic roll testing. At the same time, the testing could be conducted without the complications of installing forced oscillation equipment and the associated impact on the wind tunnel schedule.

Over the years, about 50 military and civil aircraft configurations and space vehicles have been tested at subsonic flight conditions using the free-to-roll technique, including specific and generic configurations. After decades of application, the technique has proven to be a reliable indicator of degraded or unstable roll damping. Using the technique, researchers have been able to identify aerodynamic mechanisms that caused the degraded characteristics and evaluate the effectiveness of configuration modifications designed to eliminate or minimize the problem. Many of these evaluation studies have resulted in surprising conclusions. In particular, the strong vortex flows produced by chined fuselage forebodies have played a critical role in the degradation of roll damping at high angles of attack for some fighter configurations.[3] In addition to its experimental value, the free-to-roll testing technique has resulted in worldwide studies to develop analytical and computational methods to predict the rolling motions based on the geometry of the configuration of interest.

During the development of the F/A-18E/F aircraft, an abrupt uncommanded roll-off was experienced by flight-test aircraft at transonic speeds. After the configuration was modified and entered production, NASA participated in a joint program with the Navy and Boeing to advance the state of the art in experimental and computational methods for prediction of abrupt wing stall phenomena at transonic conditions.[4] As part of the effort, NASA developed a transonic rig for the Langley 16-Foot Transonic Tunnel and applied the free-to-roll technique to four configurations: the F-16C, the F/A-18C, the F/A-18E preproduction aircraft, and the AV-8B.[5] The first two configurations are known to not exhibit wing drop or wing rock, but the last two exhibited severe wing drop transonically. In tests of all four configurations, good correlation was obtained with full-scale flight tests. The transonic free-to-roll test capability was subsequently used for joint NASA–Lockheed Martin evaluations of the F-35 configuration, with great success.[6] As a result of this test and other associated tunnel tests, the wing configuration of the F-35C for the Navy was modified to eliminate wing-drop tendencies.

The free-to-roll technique has been adapted for high-speed, high Reynolds number testing in the Langley Transonic Dynamics Tunnel, the Langley National Transonic Facility, and the Calspan Transonic Wind Tunnel.

Rotary-Balance Tests

In addition to free-spinning tests, the Spin Tunnel is equipped with an apparatus that permits measurements of six-component aerodynamic data during simulated spinning motions. Mounted to an electronic strain-gauge balance, the model is forced to spin about a vertical or inclined axis relative to the tunnel airstream at specified rates. Data extracted from the tests are used in analytical predictions of spin characteristics and for assessing the impact of configuration modifications. In general, scaled free-spinning models used in the Langley 20-Foot Spin Tunnel are too small to accommodate force and moment balances. In addition, the requirements to provide timely information on spin recovery characteristics for a large number of possible mass loadings have historically been higher priorities than interest in acquiring aerodynamic spin data with the models for most test programs. Instead, separate larger-scale models are constructed for testing on the tunnel rotary-balance test apparatus. The rotary-balance testing technique has been considerably updated and made more capable for the past 30 years. As expected, the primary interest in aerodynamic rotary testing is to provide data for the analysis of spin recovery characteristics using analytical motion calculations. In addition, the technique provides detailed data on the effect of configuration variables on aerodynamic parameters during simulated spinning motions. For example,

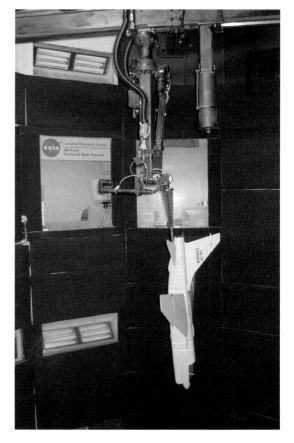

The latest rotary-balance test apparatus in the Langley Spin Tunnel permits roll oscillations to be superimposed on the steady spinning motion.

the effect of raising a horizontal tail from a low position to a T-tail location on spin damping can be readily measured and analyzed. In recent years, the rotary-balance models have become more sophisticated, including the implementation of pressure ports for measurements of pressures on wings, fuselages, and tails during spins.[7]

New developments in rotary testing at Langley include a rig that permits computer-controlled oscillatory motion superimposed on the steady spinning motions. A far cry from the archaic testing conducted in the 5-Foot Vertical Tunnel in the 1920s, this test rig provides unprecedented analysis capability for studies of the spin and recovery.

Free-to-Pitch Tests

Another unique test technique developed by NASA in the 12-Foot Low-Speed Tunnel and the Spin Tunnel to assess aerodynamic and dynamic behavior at extreme angles of attack is known as the free-to-pitch test. The technique was first stimulated by concern over deep stall and tumbling characteristics exhibited by flying wing and statically unstable configurations in the 1940s. As discussed in the previous section for X-29 tests, the test setup for free-to-pitch testing consists of mounting the model in a yoke-type apparatus, which permits a single degree of freedom in pitch with 360 degrees of rotational freedom. The yoke is placed in the vertical airstream of the Spin Tunnel, with the model attitude fixed at a prescribed value using an artificial restraint, and the model is then released and the ensuing motions observed and recorded.

Aerodynamic controls can be deflected by remote control, and the ability to terminate the tumbling motions with various combinations of control inputs can be determined. Using the technique with the model initially at extreme poststall angles of attack, researchers can evaluate whether the configuration shows susceptibility to stabilizing in an unrecoverable deep stall, or whether 360-degree tumbling motions will be induced during the recovery to conventional flight.

CHAPTER 9:

FUTURE PERSPECTIVES

O ver 80 years of conceptual studies, developmental efforts, and applications of free-flying dynamic model test techniques by the NACA and NASA have provided the aerospace community with a rich legacy of advances in the state of the art of flight dynamics. The ability to predict the behavior of full-scale vehicles for complex flight conditions has proven an ongoing challenge that has led to many technical issues that were subsequently met and eliminated. The role of dynamic models has been especially valuable in applications to the Nation's military and civil aerospace vehicles that were unconventional. As would be expected, many lessons have been learned regarding the cost-effective applications of the testing techniques and the technical limitations of free-flight models. With so much success in the rearview mirror, a personal perspective on the future opportunities and challenges to the testing techniques might be of interest to the reader.

It is appropriate to begin with an awareness of technical factors that influence the use of free-flight models for research and development. The accelerated development of remotely piloted unmanned aerial vehicles (UAVs) by the military has led to daily operations of sophisticated long-range vehicles and the attendant technologies required to develop, control, and maintain them. At the same time, a new breed of pilot has become adapted to the remote control and mission demands of the digital age asset. In comparison to today's capabilities, NASA's pioneering efforts seem almost primitive, but the Agency takes pride in its contributions, which formed a solid base for the capabilities now used so successfully by the military and industry. It is difficult to anticipate hardware breakthroughs in free-flight model technologies beyond those in evidence, but NASA's most valuable contributions have come from the applications of the models to specific aerospace issues—especially those that require years of difficult research and dedication.

It is also appropriate to observe world scenarios that are having an impact on the role of dynamically scaled models for research within NASA. Foremost among these factors is the dramatic decrease in worldwide research and development on radical military and civil aircraft since the mid-1990s. The end of the Cold War, megamergers in the aerospace industry, and the demise of funding for aeronautical research have led to a sharp reduction in research activities for unconventional vehicles. The development times required for new military aircraft has grown to decades, and the slowdown in requirements for the ongoing NASA capabilities in free-flight testing is a great concern. Adding to these concerns are the constrains implemented after the events of September 11, 2001, including those concerning the ability to fly large remotely piloted vehicles in U.S. airspace. Strict airspace control by regulatory security directives can have (and are having) an impact on some free-flight research activities, resulting in delays in model flight opportunities. Finally, planned closures of key NASA facilities will challenge new generations of researchers to reinvent the free-flight capabilities discussed herein. For example, the planned demolition of the Langley Full-Scale Tunnel after its September 2009 closure will terminate that historic facility's role in providing testing capability, although exploratory free-flight tests have been conducted in the smaller test section of the Langley 14- by 22-Foot Subsonic Tunnel. Based on the foregoing observations, NASA will be challenged to provide the facilities and expertise required to continue to provide the Nation with contributions from free-flight models.

Without doubt, the most important technical issue in the application of dynamically scaled free-flight models is the effect of Reynolds number. As discussed in previous sections, the anticipation and interpretation of scale effects will continue to be a challenge. The British Royal Aircraft Establishment (R.A.E.) had attacked the issue with the design and construction of a 15-foot pressurized spin tunnel by the National Aeronautical Establishment at Bedford, England, in the early 1950s. The approach was to provide the capability of pressurizing the tunnel to 4 atmospheres for increased Reynolds number.[1] Tests were to be conducted under pressurized conditions with the test crew removed from the tunnel for safety, and a periscope would have been used to observe the model. Launching and retrieving the model would be accomplished by means of a nylon cord attached to the model. Tests were conducted and completed with a 0.13-scale model of the tunnel to check out the design, but a fire destroyed the 15-foot pressurized spin tunnel soon after it was completed in 1954. To the author's knowledge, no other pressurized tunnel has been constructed for high-Reynolds-number capability for spin research.

In the author's opinion, the research community should seriously examine the possibilities of combining current capabilities in cryogenic wind tunnel technology, magnetic suspension systems, and other relevant fields in a feasibility study of free-spinning tests at full-scale values of Reynolds number. The obvious issues of cost, operational efficiencies, and value added versus today's testing would be critical factors in the study, although one would hope that the operational experiences gained in the U.S. and Europe with cryogenic tunnels in recent years might provide optimism for success.

In lieu of a feasible facility for full-scale spin simulation, other approaches might involve the application of computational fluid dynamics (CFD). In the 1990s, the surge of interest in emerging CFD methods resulted in a proposed joint NASA Langley–R.A.E. Bedford effort to evaluate the ability of CFD to predict Reynolds number effects for spinning bodies, such as those experienced in the XF8U-1 program mentioned earlier. The objective was of special interest in the U.S., because many wind tunnels were being closed and dismantled, including the Ames 12-foot pressure tunnel. Langley and Ames had conducted investigations of fuselage-induced Reynolds number effects and determined the need for corrective strakes for spin models. The approach used for the proposed joint study was to computationally replicate experimental data that had been generated by Edward C. Polhamus and others on Reynolds number effects of cross-sectional shapes and to use the CFL3D NASA code to compute rotary-type data for shapes that would be fabricated and tested on a rotary rig at Bedford. Unfortunately, the program was canceled before meaningful results could be obtained.

In the decade since the aborted plan, interactions between the CFD and the stability and control (S&C) communities have been limited, to the point where the CFD specialists are not aware of the needs of the S&C specialists, and the S&C specialists believe that CFD has been oversold as a replacement for wind tunnels. In recent years, a NASA effort was made (later aborted) to bring the groups together in a NASA–DOD–industry–academia program to be known as Computational Methods for Stability and Control (COMSAC).[2] One of the major thrusts of the program was to be the use of CFD to predict Reynolds number effects for low-Reynolds-number tests. Once again, this proposed project was canceled at an early stage, and no significant progress was made toward the Reynolds number issue. However, COMSAC-related research of a more limited scope has been carried forward into both the NASA Aviation Safety and Subsonic Fixed Wing programs.

In summary, the next major breakthroughs in dynamic free-flight model technology should come in the area of improving the prediction of Reynolds number effects. However, to make advances toward this goal will require a continued commitment, similar to the ones made during the past 80 years for the continued support of model testing in the areas discussed herein.

REFERENCES

Background

1. Max Scherberg and R.V. Rhode, "Mass Distribution and Performance of Free Flight Models," NACA TN-268 (1927).

2. C.H. Wolowicz, J.S. Bowman, Jr., and W.P. Gilbert, "Similitude Requirements and Scaling Relationships as Applied to Model Testing," NASA TP-1435 (1979).

3. Sanger M. Burk and Calvin F. Wilson, "Radio-Controlled Model Design and Testing Techniques for Stall/Spin Evaluation of General-Aviation Aircraft," NASA TM-80510 (1975).

4. Anshal I. Neihouse and Philip W. Pepoon, "Dynamic Similitude Between a Model and a Full-Scale Body for Model Investigation at Full-Scale Mach Number," NACA TN-2062 (1950).

5. Joseph R. Chambers and Sue B. Grafton, "Aerodynamic Characteristics of Airplanes at High Angles of Attack," NASA TM-74097 (1977).

Historical Development by the NACA and NASA

1. D. Bruce Owens, Jay M. Brandon, Mark A. Croom, Charles M. Fremaux, Eugene H. Heim, and Dan D. Vicroy, "Overview of Dynamic Test Techniques for Flight Dynamics Research at NASA LaRC," AIAA Paper 2006-3146 (2006).

2. A.V. Stephens, "Recent Research on Spinning," *Journal of the Royal Aeronautical Society* (Nov. 1933), pp. 944–955.

3. A.V. Stephens, "Free-Flight Spinning Experiments with Single-Seater Aircraft H and Bristol Fighter Models," Royal Aircraft Establishment R and M 1404 (1931).

4. H.E. Wimperis, "New Methods of Research in Aeronautics," *Journal of the Royal Aeronautical Society* (Dec. 1932), p. 985.

5. C. Wenzinger and T. Harris, "The Vertical Wind Tunnel of the National Advisory Committee for Aeronautics," NACA TR-387 (1931).

6. C.H. Zimmerman, "Preliminary Tests in the N.A.C.A. Free-Spinning Wind Tunnel," NACA TR-557 (1935).

7. Owens, et al., "Overview of Dynamic Test Techniques."

8. Robert A. Mitcheltree, Charles M. Fremaux, and Leslie A. Yates, "Subsonic Static and Dynamic Aerodynamics of Blunt Entry Vehicles," AIAA Paper 99-1020 (1999).

9. Joseph A. Shortal and Clayton J. Osterhout, "Preliminary Stability and Control Tests in the NACA Free-Flight Tunnel and Correlation with Flight Tests," NACA TN-810 (1941).

10. Alvin Seiff, Carlton S. James, Thomas N. Canning, and Alfred G. Boissevain, "The Ames Supersonic Free-Flight Wind Tunnel," NACA RM-A52A24 (1952).

11. Edwin P. Hartman, *Adventures in Research: A History of Ames Research Center 1940–1965*, NASA SP-4302 (1970).

12. Alfred G. Boissevain and Peter F. Intrieri, "Determination of Stability Derivatives from Ballistic Range Tests of Rolling Aircraft Models," NASA TM-X-399 (1961).

13. Charles J. Cornelison, "Status Report for the Hypervelocity Free-Flight Aerodynamic Facility," *48th Aero Ballistic Range Association Meeting, Austin, TX, Nov. 1997.*

14. M.C. Wilder, D.C. Reda, D.W. Bogdanoff, and D.K. Prabhu, "Free-Flight Measurements of Conductive Heat Transfer Rates in Hypersonic Ballistic-Range Environments," AIAA Paper 2007-4404 (2007).

15. Peter F. Intrieri, "Study of Stability and Drag at Mach Numbers from 4.5 to 13.5: a Topical Venus-Entry Body," NASA TN-D-2827 (1965); Tim Tam, Stephen Ruffin, Leslie Yates, John Morgenstern, Peter Gage, and David Bogdanoff, "Sonic Boom Testing of Artificially Blunted Leading Edge (ABLE) Concepts in the NASA Ames Aeroballistic Range," AIAA Paper 2000-1011 (2001).

16. Mark W. Kelly and Louis H. Smaus, "Flight Characteristics of a 1/4-Scale Model of the XFV-1 Airplane," NACA RM-SA52J15 (1952).

17. Ralph W. Stone, Jr., William G. Garner, and Lawrence J. Gale, "Study of Motion of Model of Personal-Owner or Liaison Airplane Through the Stall and Into the Incipient Spin by Means of a Free-Flight Testing Technique," NACA TN-2923 (1953).

18. Charles E. Libby and Sanger M. Burk, Jr., "A Technique Utilizing Free-Flying Radio-Controlled Models to Study the Incipient-and Developed-Spin Characteristics of Airplanes," NASA Memo 2-6-59L (1959).

19. David J. Fratello, Mark A. Croom, Luat T. Nguyen, and Christopher S. Domack, "Use of the Updated NASA Langley Radio-Controlled Drop-Model Technique for High-Alpha Studies of the X-29A Configuration," AIAA Paper 1987-2559 (1987).

20. Mark A. Croom, Holly M. Kenney, and Daniel G. Murri, "Research on the F/A-18E/F Using a 22%-Dynamically-Scaled Drop Model," AIAA Paper 2000-3913 (2000).

21. R. Dale Reed, *Wingless Flight: The Lifting Body Story*, NASA SP-4220 (1997).

22. Euclid C. Holleman, "Summary of Flight Tests to Determine the Spin and Controllability Characteristics of a Remotely Piloted, Large-Scale (3/8) Fighter Airplane Model," NASA TN-D-8052 (1976).

23. Staff of the Langley Research Center, "Exploratory Study of the Effects of Wing-Leading-Edge Modifications on the Stall/Spin Behavior of a Light General Aviation Airplane," NASA TP-1589 (1979).

24. Kevin Cunningham, John V. Foster, Eugene A. Morelli, and Austin M. Murch, "Practical Application of a Subscale Transport Aircraft for Flight Research in Control Upset and Failure Conditions," AIAA Paper 2008-6200 (2008).

25. Robert O. Schade, "Flight Test Investigation on the Langley Control-Line Facility of a Model of a Propeller-Driven Tail-Sitter-Type Vertical-Take-off Airplane with Delta Wing During Rapid Transitions," NACA TN-4070 (1957).

26. Michael J. Hirschberg and David M. Hart, "A Summary of a Half-Century of Oblique Wing Research," AIAA Paper 2007-150 (2007).

27. D.A. Deets and L.E. Brown, "Wright Brothers Lectureship in Aeronautics: Experience with HiMAT Remotely Piloted Research Vehicle-An Alternate Flight Test Approach," AIAA Paper 86-2754 (1986).

28. Laurence A. Walker, "Flight Testing the X-36—The Test Pilot's Perspective," NASA CR-198058 (1997).

29. Joseph A. Shortal, *A New Dimension: Wallops Island Flight Test Range, the First Fifteen Years*, NASA RP-1028 (1978).

Selected Applications

1. Scherberg and Rhode, "Mass Distribution and Performance of Free Flight Models," NACA TN-268.

2. Joseph R. Chambers, *Partners in Freedom: Contributions of the NASA Langley Research Center to Military Aircraft of the 1990s*, NASA SP-2000-4519 (2000).

3. Malcolm J. Abzug and E. Eugene Larrabee, *Airplane Stability and Control* (Cambridge University Press, 2002).

Dynamic Stability and Control

1. Shortal and Osterhout, "Preliminary Stability and Control Tests in the NACA Free-Flight Tunnel."

2. Herman O. Ankenbruck, "Determination of the Stability and Control Characteristics of a Straight-Wing, Tailless Fighter-Airplane Model in the Langley Free-Flight Tunnel," NACA Wartime Report ACR No. L5K05 (1946).

3. Fred E. Weick and Carl J. Wenzinger, "Preliminary Investigation of Rolling Moments Obtained with Spoilers on Both Slotted and Plain Wings," NACA TN-415 (1932).

4. M.O. McKinney, "Experimental Determination of the Effects of Dihedral, Vertical Tail Area, and Lift Coefficient on Lateral Stability and Control Characteristics," NACA TN-1094 (1946).

5. John P. Campbell and Charles L. Seacord, Jr., "The Effect of Mass Distribution on the Lateral Stability and Control Characteristics of an Airplane as Determined by Tests of a Model in the Free-Flight Tunnel," NACA TR-769 (1943).

6. C.C. Smith, "The Effects of Fuel Sloshing on the Lateral Stability of a Free-Flying Airplane Model," NACA RM-L8C16 (1948).

7. William R. Bates, "Static Stability of Fuselages Having a Relatively Flat Cross Section," NACA TN-3429 (1955).

8. Robert O. Schade and James L. Hassell, Jr., "The Effects on Dynamic Lateral Stability and Control of Large Artificial Variations in the Rotary Stability Derivatives," NACA TN-2781 (1953).

9. Shortal, *A New Dimension*.

10. Carl A. Sandahl, "Free-Flight Investigation at Transonic and Supersonic Speeds of a Wing-Aileron Configuration Simulating the D558-2 Airplane," NACA RM-L8E28 (1948); Carl A. Sandahl, "Free-Flight Investigation at Transonic and Supersonic Speeds of the Rolling Effectiveness for a 42.7° Sweptback Wing Having Partial-Span Ailerons," NACA RM-L8E25 (1948).

11. Examples include James H. Parks and Jesse L. Mitchell, "Longitudinal Trim and Drag Characteristics of Rocket-Propelled Models Representing Two Airplane Configurations," NACA RM-L9L22 (1949); and James L. Edmondson and E. Claude Sanders, Jr., "A Free-Flight Technique for Measuring Damping in Roll by Use of Rocket-Powered Models and Some Initial Results for Rectangular Wings," NACA RM-L9I01 (1949).

12. James H. Parks, "Experimental Evidence of Sustained Coupled Longitudinal and Lateral Oscillations From Rocket-Propelled Model of a 35° Swept-Wing Airplane Configuration," NACA RM-L54D15 (1954).

13. Grady L. Mitcham, Joseph E. Stevens, and Harry P. Norris, "Aerodynamic Characteristics and Flying Qualities of a Tailless Triangular-Wing Airplane Configuration as Obtained from Flights of Rocket-Propelled Models at Transonic and Low Supersonic Speeds," NACA TN-3753 (1956).

14. William N. Gardner, "Aerodynamic Characteristics at Transonic Speeds of a 1/7-Scale Rocket-Powered Model of the Grumman XF10F Airplane With Wing Sweepback of 42.5°," NACA RM-SL54D16 (1954).

15. John V. Becker, *The High Speed Frontier*, NASA SP-445 (1980).

16. Maurice L. Rasmussen, "Determination of Nonlinear Pitching-Moment Characteristics of Axially Symmetric Models From Free-Flight Data," NASA TN-D-144 (1960).

17. Alfred G. Boissevain and Peter F. Intrieri, "Determination of Stability Derivatives from Ballistic Range Tests of Rolling Aircraft Models," NASA TM-X-399 (1961).

18. Joseph R. Chambers and Mark A. Chambers, *Radical Wings and Wind Tunnels* (Specialty Press, 2008).

19. Charles Donlan, "An Interim Report on the Stability and Control of Tailless Airplanes," NACA TR-796 (1944).

20. J.P. Campbell and M.O. McKinney, "Summary of Methods for Calculating Dynamic Lateral Stability and Response and for Estimating Lateral Stability Derivatives," NACA TN-2409 (1951).

21. P.C. Boisseau, "Investigation in the Langley Free-Flight Tunnel of the Low-Speed Stability and Control Characteristics of a 1/10-Scale Model Simulating the Convair F-102A Airplane," NACA RM-SL55B21 (1955).

22. Charles J. Donlan and William C. Sleeman, "Low-Speed Wind-Tunnel Investigation of the Longitudinal Stability Characteristics of a Model Equipped with a Variable-Sweep Wing," NACA RM-L9B18 (1949).

23. Forward-swept wing studies included: C.V. Bennett, "Investigation of the Stability and Control Characteristics of a 1/20-Scale Model of the Consolidated Vultee XB-53 Airplane in the Langley Free-Flight Tunnel," NACA RM-SL7J17 (1947); B. Maggin and A.H. LeShane, "Tow Tests of a 1/17.8-Scale Model of the XFG-1 Glider in the Langley Free-Flight Tunnel," NACA MR-L5H21 (1945); and H.O. Ankenbruck, "Determination of the Stability and Control Characteristics of a 1/10-Scale Model of the MCD-387-A Swept-Forward Wing, Tailless Fighter Airplane in the Langley Free-Flight Tunnel," NACA MR-L6G02 (1946).

24. Joseph L. Johnson, "Stability and Control Characteristics of a 1/10-Scale Model of the McDonnell XP-85 Airplane While Attached to the Trapeze," NACA RM-L7J16 (1947).

25. Robert E. Shanks and David C. Grana, "Flight Tests of a Model Having Self-Supporting Fuel-Carrying Panels Hinged to the Wing Tips," NACA RM-L9107a (1949); C.V. Bennett and P.C. Boisseau, "Free-Flight-Tunnel Investigation of the Dynamic Lateral Stability and Control Characteristics of a High-Aspect-Ratio Bomber Model with a Sweptback-Wing Fighter Model Attached to Each Wing Tip," NACA RM-L52E03 (1952).

26. Chambers, *Radical Wings and Wind Tunnels*.

27. William R. Bates, Powell M. Lovell, Jr., and Charles C. Smith, Jr., "Dynamic Stability and Control Characteristics of a Vertically Rising Airplane Model in Hovering Flight," NACA RM-L50J16 (1951).

28. Hovering and transition tests included: P.M. Lovell, C.C. Smith, and R.H. Kirby, "Stability and Control Flight Tests of a 0.13-Scale Model of the Consolidated Vultee XFY-1 Airplane in Take-Offs, Landings, and Hovering Flight," NACA RM-SL52I26 (1952); and P.M. Lovell, C.C. Smith, and R.H. Kirby, "Flight Investigation of the Stability and Control Characteristics of a 0.13-Scale Model of the Convair XFY-1 Vertically Rising Airplane During Constant-Altitude Transitions," NACA RM-SL53E18 (1953).

29. Robert O. Schade, "Flight Test Investigation on the Langley Control-Line Facility of a Model of a Propeller-Driven Tail-Sitter-Type Vertical-Take-off Airplane with Delta Wing During Rapid Transitions," NACA TN-4070 (1957).

30. Charles C. Smith, Jr., "Hovering and Transition Flight Tests of a 1/5-Scale Model of a Jet-Powered Vertical-Attitude VTOL Research Airplane," NASA TN-D-404 (1961); Charles C. Smith, Jr., "Flight Tests of a 1/6-Scale Model of the Hawker P.1127 Jet VTOL Airplane," NASA TM-SX-531 (1961); Louis P. Tosti, "Rapid-Transition Tests of a ¼-Scale Model of the VZ-2 Tilt-Wing Aircraft," NASA TN-D-946 (1961).

31. Robert H. Kirby, "Stability and Control Flight Tests of a Vertically Rising Airplane Model Similar to the Lockheed XFV-1 Airplane," NACA RM-SL54J18 (1954); Robert H. Kirby, "Flight Investigation of the Stability and Control characteristics of a Vertically Rising Airplane with Swept or Unswept Wings and x- or +- Tails," NACA TN-3812 (1956).

32. Mark W. Kelly and Lewis H. Smaus, "Flight Characteristics of a 1/4-Scale Model of the XFV-1 Airplane," NACA RM-SA52J15 (1952).

33. W.A. Newsom, Jr., and E.L. Anglin, "Free-Flight Model Investigation of a Vertical Attitude VTOL Fighter," NASA TN-D-8054 (1975); Sue B. Grafton and Ernie L. Anglin, "Free-Flight Model Investigation of a Vertical-Attitude VTOL Fighter with Twin Vertical Tails," NASA TN-D-8089 (1975).

34. Robert Kress, "The Last Frontier: VTOL," *Flight Journal Magazine* (June 2003).

35. Chambers, *Partners in Freedom*.

36. Smith, "Flight Tests of the Hawker P.1127."

37. Powell M. Lovell, Jr., and Lysle P. Parlett, "Hovering-Flight Tests of a Model of a Transport Vertical Take-Off Airplane with Tilting Wing and Propellers," NACA TN-3630 (1956); Powell M. Lovell, Jr., and Lysle P. Parlett, "Flight Tests of a Model of a High-Wing Transport Vertical-Take-Off Airplane With Tilting Wing and Propellers and With Jet Controls at the Rear of the Fuselage for Pitch and Yaw Control," NACA TN-3912 (1957).

38. Louis P. Tosti, "Flight Investigation of Stability and Control Characteristics of a 1/8-Scale Model of a Tilt-Wing Vertical-Take-Off-And-Landing Airplane," NASA TN-D-45 (1960); Louis P. Tosti, "Longitudinal Stability and Control of a Tilt-Wing VTOL Aircraft Model with Rigid and Flapping Propeller Blades," NASA TN-D-1365 (1962); William A. Newsom and Robert H. Kirby, "Flight Investigation of Stability and Control Characteristics of a 1/9-Scale Model of a Four-Propeller Tilt-Wing V/STOL Transport," NASA TN-D-2443 (1964).

39. John P. Campbell, Jr., and Joseph L. Johnson, Jr., "Wind-Tunnel Investigation of an External-Flow Jet-Augmented Slotted Flap Suitable for Applications to Airplanes with Pod-Mounted Jet Engines," NACA TN-3898 (1956).

40. Lysle P. Parlett, "Free-Flight Wind-Tunnel Investigation of a Four-Engine Sweptwing Upper-Surface Blown Transport Configuration," NASA TM-X-71932 (1974); Lysle P. Parlett, "Free-Flight Investigation of the Stability and Control Characteristics of a STOL Model with an Externally Blown Jet Flap," NASA TN-D-7411 (1974); Chambers, *Radical Wings and Wind Tunnels*.

41. Donald E. Hewes, "Free-Flight Investigation of Radio-Controlled Models with Parawings," NASA TN-D-927 (1961).

42. Joseph L. Johnson, Jr., "Low-Speed Wind-Tunnel Investigation to Determine the Flight Characteristics of a Model of a Parawing Utility Vehicle," NASA TN-D-1255 (1962).

43. Joseph L. Johnson, Jr., "Low-Speed Force and Flight Investigation of a Model of a Modified Parawing Utility Vehicle," NASA TN-D-2492 (1965).

44. Sanger M. Burk, Jr., "Free-Flight Investigation of the Deployment, Dynamic Stability, and Control Characteristics of a 1/12-Scale Dynamic Radio-Controlled Model of a Large Booster and Parawing," NASA TN-D-1932 (1963).

45. Charles E. Libbey, "Free-Flight Investigation of the Deployment of a Parawing Recovery Device for a Radio-Controlled 1/5-Scale Dynamic Model Spacecraft," NASA TN-D-2044 (1963).

46. Chambers, *Radical Wings and Wind Tunnels*.

47. Chambers, *Partners in Freedom*.

48. Ibid.

49. John P. Campbell and Hubert M. Drake, "Investigation of Stability and Control Characteristics of an Airplane Model with Skewed Wing in the Langley Free-Flight Tunnel," NACA TN-1208 (1947).

50. Edward C. Polhamus and Thomas A. Toll, "Research Related to Variable Sweep Aircraft Development," NASA TM-83121 (1981).

51. Michael J. Hirschberg and David M. Hart, "A Summary of a Half-Century of Oblique Wing Research," AIAA Paper 2007-150 (2007).

52. Ibid.

53. James E. Murray, Joseph W. Pahle, Stephen P. Thornton, Shannon Vogus, Tony Frackowiak, Joe Mello, and Brook Norton, "Ground and Flight Evaluation of a Small-Scale Inflatable-Winged Aircraft," NASA TM-2002-210721 (2002).

54. Chambers, *Radical Wings and Wind Tunnels*.

55. Joel S. Levine, et al., "Science from a Mars Airplane: The Aerial Regional-Scale Environmental Survey (ARES) of Mars," AIAA Paper 2003-6576 (2003); P. Sean Kenney and Mark A. Croom, "Simulating the ARES Aircraft in the Mars Environment," AIAA Paper 2003-6579 (2003).

56. James S. Bowman, Jr., "Dynamic Model Tests at Low Subsonic Speeds of Project Mercury Capsule Configurations With and Without Drogue Parachutes," NASA TM-X-459 (1961); Henry A. Lee, Peter S. Costigan, and James S. Bowman, Jr., "Dynamic Model Investigation of a 1/20-Scale Gemini Spacecraft in the Langley Spin Tunnel," NASA TN-D-2191 (1964); Henry A. Lee and Sanger M. Burk, "Low-Speed Dynamic Model Investigation of Apollo Command Module Configuration in the Langley Spin Tunnel," NASA TN-D-3888 (1967).

57. Peter S. Costigan, "Dynamic-Model Study of Planetary-Entry Configurations in the Langley Spin Tunnel," NASA TN-D-3499 (1966).

58. Robert A. Mitcheltree, Charles M. Fremaux, and Leslie A. Yates, "Subsonic Static and Dynamic Aerodynamics of Blunt Entry Vehicles."

59. Charles M. Fremaux and R. Keith Johnson, "Subsonic Dynamic Stability Tests of a Sample Return Entry Vehicle," *4th International Planetary Probe Workshop, Pasadena, CA, June 2006*.

60. David E. Hahne and Charles M. Fremaux, "Low-Speed Dynamic Tests and Analysis of the Orion Crew Module Drogue Parachute System," AIAA Paper 2008-09-05 (2008).

61. Ibid.

62. Peter C. Boisseau, "Investigation of the Low-Speed Stability and Control Characteristics of a 1/7-Scale Model of the North American X-15 Airplane," NACA RM-L57D09 (1957); Donald E. Hewes and James L. Hassell, Jr., "Subsonic Flight Tests of a 1/7-Scale Radio-Controlled Model of the North American X-15 Airplane With Particular Reference to High Angle-of-Attack Conditions," NASA TM-X-283 (1960).

63. Dennis R. Jenkins and Tony R. Landis, *Hypersonic-The Story of the North American X-15* (Specialty Press, 2008).

64. John W. Paulson, Robert E. Shanks, and Joseph L. Johnson, "Low-Speed Flight Characteristics of Reentry Vehicles of the Glide-Landing Type," NASA TM-X-331 (1960).

65. Robert E. Shanks and George M. Ware, "Investigation of the Flight Characteristics of a 1/5-Scale Model of a Dyna-Soar Glider Configuration at Low Subsonic Speeds," NASA TM-X-683 (1962).

66. James L. Hassell, Jr., "Investigation of the Low-Subsonic Stability and Control Characteristics of a 1/3-Scale Free-Flying Model of a Lifting-Body Reentry Configuration," NASA TM-X-297 (1960).

67. R. Dale Reed, *Wingless Flight: The Lifting Body Story*, NASA SP-4220 (1997).

68. Robert E. Shanks, "Investigation of the Dynamic Stability and Controllability of a Towed Model of a Modified Half-Cone Reentry Vehicle," NASA TN-D-2517 (1965).

69. George M. Ware, "Investigation of the Flight Characteristics of a Model of the HL-10 Manned Lifting Entry Vehicle," NASA TM-X-1307 (1967).

70. Robert E. Shanks, "Investigation of the Dynamic Stability of Two Towed Models of a Flat-Bottom Lifting Reentry Configuration," NASA TM-X-1150 (1966).

71. Reed, *Wingless Flight*.

72. Alex G. Sim, James E. Murray, David C. Neufeld, and R. Dale Reed, "The Development and Flight Test of a Deployable Precision Landing System for Spacecraft Recovery," NASA TM-4525 (1993).

73. Delma C. Freeman, Jr., "Low Subsonic Flight and Force Investigation of a Supersonic Transport Model with a Variable-Sweep Wing," NASA TN-D-4726 (1968); Delma C. Freeman, Jr., "Low Subsonic Flight and Force Investigation of a Supersonic Transport Model with a Double-Delta Wing," NASA TN-D-4179 (1968).

74. Delma C. Freeman, Jr., "Low Subsonic Flight and Force Investigation of a Supersonic Transport Model with a Highly Swept Arrow Wing," NASA TN-D-3887 (1967).

75. Linwood W. McKinney and Edward C. Polhamus, "A Summary of NASA Data Relative to External-Store Separation Characteristics," NASA TN-D-3582 (1966).

76. William J. Alford, Jr., and Robert T. Taylor, "Aerodynamic Characteristics of the X-15/B-52 Combination," NASA Memo 8-59L (1958).

77. Robert W. Rainey, "A Wind-Tunnel Investigation of Bomb Release at a Mach Number of 1.62," NACA RM-L53L29 (1954).

78. Shortal, *A New Dimension.*

79. Carl A. Sandahl and Maxime A. Faget, "Similitude Relations for Free-Model Wind-Tunnel Studies of Store-Dropping Problems," NACA TN-3907 (1957).

80. Chambers, *Radical Wings and Wind Tunnels.*

81. Dan D. Vicroy, "Blended-Wing-Body Low-Speed Flight Dynamics: Summary of Ground Tests and Sample Results," AIAA Invited Paper Presented at the *47th AIAA Aerospace Sciences Meeting and Exhibit, Jan. 2009.*

82. Joseph R. Chambers, *Concept to Reality*, NASA SP 2003-4529 (2003).

83. Dale R. Satran, "Wind-Tunnel Investigation of the Flight Characteristics of a Canard General-Aviation Airplane Configuration," NASA TP-2623 (1986).

84. Jay M. Brandon, Frank L. Jordan, Jr., Robert A. Stuever, and Catherine W. Buttrill, "Application of Wind Tunnel Free-Flight Technique for Wake Vortex Encounters," NASA TP-3672 (1997).

85. Robert E. Shanks, "Aerodynamic and Hydrodynamic Characteristics of Models of Some Aircraft-Towed Mine-Sweeping Devices," NACA RM-SL55K21 (1955); Robert E. Shanks, "Experimental Investigation of the Dynamic Stability of a Towed Parawing Glider Model," NASA TN-D-1614 (1963); Robert E. Shanks, "Investigation of the Dynamic Stability and Controllability of a Towed Model of a Modified Half-Cone Reentry Vehicle," NASA TN-D-2517 (1965).

86. Marvin Pitkin and Robert E. Shanks, "Flight Tests of a Glider Model Towed by Twin Parallel Towlines," NACA RB-3D30 (1943).

87. B. Maggin and A.H. LeShane, "Tow Tests of a 1/17.8-Scale Model of the XFG-1 Glider in the Langley Free-Flight Tunnel," NACA MR-L5H21 (1945); B. Maggin and C.V. Bennett, "Tow Tests of a 1/17-Scale Model of the P-80A Airplane in the Langley Free-Flight Tunnel," NACA MR-L5K06 (1945).

88. B. Maggin and Robert E. Shanks, "Experimental Determination of the Lateral Stability of a Glider Towed by a Single Towline and Correlation with an Approximate Theory," NACA RM-L8H23 (1948).

89. Robert E. Shanks, "Free-Flight Tunnel Investigation of the Stability and Control of a Republic F-84E Airplane Towed by a Short Towline," NACA RM-SL52K13a (1952); D.C. Grana and R.E. Shanks,

"Free-Flight Tunnel Investigation of the Dynamic Stability and Control Characteristics of a Chance Vought F7U-3 Airplane in Towed Flight," NACA RM-SL53D07 (1953).

90. Dwain A. Deets and Carl A. Crother, "Highly Maneuverable Aircraft Technology," *AGARD Active Controls in Aircraft Design* (1978).

91. Dwain A. Deets, V. Michael DeAngelis, and David P. Lux, "HiMAT Flight Program: Test Results and Program Assessment Overview," NASA TM-86725 (1986).

92. Laurence A. Walker, "Flight Testing the X-36-The Test Pilot's Perspective," NASA CR-198058 (1997).

Flight at High Angles of Attack

1. John P. Campbell, "Free and Semi-Free Model Flight-Testing Techniques Used in Low-Speed Studies of Dynamic Stability and Control," NATO Advisory Group for Aeronautical Research and Development AGARDograph 76 (1963).

2. John W. Draper, "An Investigation of the Low-Speed Stability and Control Characteristics of a 1/10-Scale Model of the McDonnell XF3H-1 Airplane," NACA RM-SL51J12 (1951).

3. Marion O. McKinney, Jr., and Hubert M. Drake, "Flight Characteristics at Low Speed of Delta-Wing Models," NACA RM-L7K07 (1948).

4. Joseph L. Johnson, Jr., "Investigation of the Low-Speed Stability and Control Characteristics of a 1/10-Scale Model of the Douglas XF4D-1 Airplane in the Langley Free-Flight Tunnel," NACA RM-SL51J22 (1951); Peter C. Boisseau, "Investigation in the Langley Free-Flight Tunnel of the Low-Speed Stability and Control Characteristics of a 1/10-Scale Model Simulating the Convair F-102A Airplane," NACA RM-SL55B21 (1955).

5. Joseph R. Chambers and Ernie L. Anglin, "Analysis of Lateral Directional Stability Characteristics of a Twin Jet Fighter Airplane at High Angles of Attack," NASA TN-D-5361 (1969); W.A. Newsom, Jr., and S.B. Grafton, "Free-Flight Investigation of Effects of Slats on Lateral-Directional Stability of a 0.13-Scale Model of the F-4E Airplane" NASA TM-SX-2337 (1971).

6. Martin T. Moul and John W. Paulson, "Dynamic Lateral Behavior of High-Performance Aircraft," NACA RM-L58E16 (1958).

7. John W. Paulson, "Investigation of the Low-Speed Flight Characteristics of a 1/15-Scale Model of the Convair XB-58 Airplane," NACA RM-SL57K19 (1957).

8. James S. Bowman, Jr., "Spin-Entry Characteristics of a Delta-Wing Airplane as Determined by a Dynamic Model," NASA TN-D-2656 (1965).

9. John W. Paulson, Robert E. Shanks, and Joseph L. Johnson, Jr., "Low-Speed Flight Characteristics of Reentry Vehicles of the Glide-Landing Type," NASA TM-X-331 (1960).

10. L.T. Nguyen, L.P. Yip, and J.R. Chambers, "Self-Induced Wing Rock of Slender Delta Wings," AIAA Paper 81-1883 (1981).

11. Chambers, *Partners in Freedom*; *Aeronautical System Division/Air Force Flight Dynamics Laboratory Symposium on Stall/Post-Stall/Spin. Symposium held at Wright-Patterson Air Force Base, OH, Dec. 15–17, 1971.*

12. Newsom and Grafton, "Free-Flight Investigation of Effects of Slats," NASA TM-SX-2337; S.B. Grafton, J.R. Chambers, and P.L. Coe, "Wind-Tunnel Free-Flight Investigation of a Model of a Spin-Resistant Fighter Configuration," NASA TN-D-7716 (1974); P.C. Boisseau and J.R. Chambers, "Lateral-Directional Characteristics of a 1/10-Scale Free-Flight Model of a Variable-Sweep Fighter Airplane at High Angles of Attack," NASA TM-SX-2649 (1972); William P. Gilbert, "Free-Flight Investigation of Lateral-Directional Characteristics of a 0.10-Scale Model of the F-15 Airplane at High Angles of Attack," NASA TM-SX-2807 (1973); Peter C. Boisseau, "Flight Investigation of Dynamic Stability and Control Characteristics of a 1/10-Scale Model of a Variable-Wing-Sweep Fighter Airplane Configuration," NASA TM-X-1185 (1965); W.A. Newsom, Jr., and Ernie L. Anglin, "Free-Flight Investigation of a 0.15-Scale Model of the XFV-12A Airplane in Clean Configuration at High Angles of Attack," NASA TM-SX-3157 (1975); W.A. Newsom, Jr., Ernie L. Anglin, and S.B. Grafton, "Free-Flight Investigation of a 0.15-Scale Model of the YF-16 Airplane at High Angle of Attack," NASA TM-SX-3279 (1975); Sue B. Grafton, Ernie L. Anglin, and W.A. Newsom, Jr., "Free-Flight Investigation of a 0.15-Scale Model of the YF-17 Airplane at High Angles of Attack," NASA TM-SX-3217 (1975); W.A. Newsom, Jr., and Sue B. Grafton, "Free-Flight Investigation of a 1/17-Scale Model of the B-1 Airplane at High Angles of Attack," NASA TM-SX-2744 (1973); Peter C. Boisseau, "Flight Investigation of Dynamic Stability and Control Characteristics of a 1/10-Scale Model of a Variable-Wing-Sweep Fighter Airplane Configuration," NASA TM-X-1367 (1967); Daniel G. Murri, Luat T. Nguyen, and Sue B. Grafton, "Wind-Tunnel Free-Flight Investigation of a Model of a Forward-Swept Wing Fighter Configuration," NASA TP-2230 (1984); Sue B. Grafton and Luat T. Nguyen, "Wind-Tunnel Free-Flight Investigation of a Model of a Cranked-Arrow-Wing Fighter Configuration," NASA TP-2410 (1985); Frank L. Jordan, Jr., and David E. Hahne, "Wind-Tunnel Static and Free-Flight Investigation of High-Angle-of-Attack Stability and Control Characteristics of a Model of the EA-6B Airplane," NASA TP- 3194 (1992).

13. Chambers and Anglin, "Analysis of Lateral Directional Stability Characteristics of a Twin Jet Fighter Airplane," NASA TN-D-5361.

14. Chambers, *Partners in Freedom*.

15. Ibid.

16. Ibid.

17. Albion H. Bowers, et al., "An Overview of the NASA F-18 High Alpha Research Vehicle," NASA CP-1998-207676 (1998).

18. Daniel G. Murri, Gautam H. Shah, Daniel J. DiCarlo, and Todd W. Trilling, "Actuated Forebody Strake Controls for the F-18 High-Alpha Research Vehicle," *Journal of Aircraft*, vol. 32, No.3 (1995), pp. 555–562.

19. Chambers, *Partners in Freedom*.

Spinning and Spin Recovery

1. Neihouse, Klinar, and Scher, "Status of Spin Research" NASA TR-R-57.

2. Zimmerman, "Preliminary Tests in the N.A.C.A. Free-Spinning Wind Tunnel," NACA TR-557.

3. Oscar Seidman and Anshal I. Neihouse, "Free-Spinning Wind-Tunnel Tests on a Low-Wing Monoplane with Systematic Changes in Wings and Tails III. Mass Distributed Along the Wings," NACA TN-664 (1938).

4. Oscar Seidman and Charles J. Donlan, "An Approximate Spin Design Criteria for Monoplanes," NACA TN-711 (1939).

5. Anshal I. Neihouse, Jacob H. Lichtenstein, and Philip W. Pepoon, "Tail-Design Requirements for Satisfactory Spin Recovery," NACA TN-1045 (1946).

6. Ralph W. Stone, Jr., Sanger M. Burk, Jr., and William Bihrle, Jr., "The Aerodynamic Forces and Moments on a 1/10-Scale Model of a Fighter Airplane in Spinning Attitudes as Measured on a Rotary Balance in the Langley 20-Foot Free-Spinning Tunnel," NACA TN-2181 (1950).

7. Ralph W. Stone, Jr., and Walter J. Klinar, "The Influence of Very Heavy Fuselage Mass Loadings and Long Nose Lengths Upon Oscillations in the Spin," NACA TN-1510 (1948).

8. H.B. Irving, A.S. Batson, and J.H. Warsap, "The Contribution of the Body and Tail of an Aero Plane to the Yawing Moment in a Spin," British R&M No. 1689 (1936).

9. Walter J. Klinar, Henry A. Lee, and L. Faye Wilkes, "Free-Spinning-Tunnel Investigation of a 1/25-Scale Model of the Chance Vought XF8U-1 Airplane," NACA RM-SL56L31b (1956).

10. M.H. Clarkson, "Autorotation of Fuselages," *Aeronautical Engineering Review*, vol. 17 (Feb. 1958); Edward C. Polhamus, "Effect of Flow Incidence and Reynolds Number on Low-Speed Aerodynamic Characteristics of Several Noncircular Cylinders with Applications to Directional Stability and Spinning," NACA TN-4176 (1958).

11. D.N. Petroff, S.H. Scher, and L.E. Cohen, "Low Speed Aerodynamic Characteristics of an 0.075-Scale F-15 Airplane Model at High Angles of Attack and Sideslip," NASA TM-X-62360 (1974); D.N. Petroff, S.H. Scher, and C.E. Sutton, "Low-Speed Aerodynamic Characteristics of a 0.08-Scale YF-17 Airplane Model at High Angles of Attack and Sideslip," NASA TM-78438 (1978); Raymond D. Whipple and J.L. Ricket,

"Low-Speed Aerodynamic Characteristics of a 1/8-scale X-29A Airplane Model at High Angles of Attack and Sideslip," NASA TM-87722 (1986).

12. Henry A. Lee and Frederick M. Healy, "Free-Spinning-Tunnel Investigation of a 1/25-Scale Model of the Chance Vought F8U-1P Airplane," NASA TM-SX-196 (1959).

13. Stanley H. Scher and William L. White, "Spin-Tunnel Investigation of the Northrop F-5E Airplane," NASA TM-SX-3556 (1977); C. Michael Fremaux, "Wind-Tunnel Parametric Investigation of Forebody Devices for Correcting Low Reynolds Number Aerodynamic Characteristics at Spinning Attitudes," NASA CR-198321 (1996).

14. William Letko, "A Low-Speed Experimental Study of the Directional Characteristics of a Sharp-Nosed Fuselage Through a Large Angle-of-Attack Range at Zero Angle of Sideslip," NACA TN-2911 (1953); P.L. Coe, Jr., J.R. Chambers, and W. Letko, "Asymmetric Lateral-Directional Characteristics of Pointed Bodies of Revolution at High Angles of Attack," NASA TN-D-7095 (1972).

15. Ernie L. Anglin, James S. Bowman, Jr., and Joseph R. Chambers, "Effects of a Pointed Nose on Spin Characteristics of a Fighter Airplane Model Including Correlation with Theoretical Calculations," NASA TN-D-5921 (1970).

16. E.R. Keener, G.T. Chapman, L. Cohen, and J. Taleghani, "Side Forces on a Tangent Ogive Forebody with a Fineness Ratio of 3.5 at High Angles of Attack and Mach Numbers From 0.1 to 0.7," NASA TM-X-3437 (1977).

17. Neihouse, "Status of Spin Research," NASA TR-R-57.

18. James S. Bowman, Jr., "Free-Spinning-Tunnel Investigation of Gyroscopic Effects of Jet-Engine Rotating Parts (or of Rotating Propellers) on Spin and Spin Recovery," NACA TN-3480 (1955).

19. Stanley H. Scher, "Wind-Tunnel Investigation of the Behavior of Parachutes in Close Proximity to One Another," NACA RM-L53G07 (1953); Stanley H. Scher and John W. Draper, "The Effects of Stability of Spin-Recovery Tail Parachutes on the Behavior of Airplanes in Gliding Flight and in Spins," NACA TN-2098 (1950); Sanger M. Burk, Jr., "Summary of Design Considerations for Airplane Spin-Recovery Parachute Systems," NASA TN-D-6866 (1972); H. Paul Stough, III, "A Summary of Spin-Recovery Parachute Experience on Light Airplanes," AIAA Paper 90-1317 (1990).

20. Neihouse, "Status of Spin Research," NASA TR-R-57; Sanger M. Burk, Jr., and Frederick M. Healy, "Comparison of Model and Full-Scale Spin Recoveries Obtained by Use of Rockets," NACA TN-3068 (1954); Raymond D. Whipple, "Rockets for Spin Recovery," NASA CR-159240 (1980).

21. Henry A. Lee, "Spin-Tunnel Investigation of a 1/20-Scale Model of a Straight-Wing, Twin-Boom, Counter-Insurgency Airplane," NASA TM-X-1602 (1969).

22. Whipple, "Rockets for Spin Recovery," NASA CR-159240.

23. L.T. Nguyen, W.P. Gilbert, J. Gera, K.W. Illiff, and E.K. Enevoldson, "Application of High-Alpha Control System Concepts to a Variable-Sweep Fighter Airplane," AIAA Paper 80-1582 (1980).

24. James S. Bowman, Jr., and Sanger M. Burk, Jr., "Stall/Spin Studies Relating to Light General-Aviation Aircraft," SAE Paper Presented at the *Society of Automotive Engineers Business Aircraft Meeting, Wichita, KS, Apr. 1973*.

25. Sanger M. Burk, Jr., James S. Bowman, Jr., and William L. White, "Spin-Tunnel Investigation of the Spinning Characteristics of Typical Single-Engine General Aviation Airplane Designs: Part I-Low-Wing Model A.: Effects of Tail Configurations," NASA TP-1009 (1977).

26. H. Paul Stough, III, James S. Patton, Jr., and Steven M. Sliwa, "Flight Investigation of the Effect of Tail Configuration on Stall, Spin, and Recovery Characteristics of a Low-Wing General Aviation Research Airplane," NASA TP-2644 (1987).

27. H. Paul Stough, III, "A Summary of Spin-Recovery Parachute Experience on Light Airplanes," AIAA Paper 90-1317 (1990).

28. M.L. Holcomb, "The Beech Model 77 "Skipper" Spin Program," AIAA Paper 79-1835 (1979).

29. William L. White and James S. Bowman, Jr., "Spin-Tunnel Investigation of a 1/13-Scale Model of the NASA AD-1 Oblique-Wing Research Aircraft," NASA TM-83236 (1982).

Spin Entry and Poststall Motions

1. Ralph W. Stone, Jr., William G. Garner, and Lawrence J. Gale, "Study of Motion of Model of Personal-Owner or Liaison Airplane Through the Stall and into the Incipient Spin by Means of a Free-Flight Testing Technique," NACA TN-2923 (1953).

2. Libby, "A Technique Utilizing Free-Flying Radio-Controlled Models," NASA Memo 2-6-59L.

3. Donald E. Hewes and James L. Hassell, Jr., "Subsonic Flight Test of a 1/7-Scale Radio-Controlled Model of the North American X-15 Airplane with Particular Reference to High Angle-of-Attack Conditions," NASA TM-X-283 (1960).

4. Sanger M. Burk, Jr., and Charles E. Libby, "Large-Angle Motion Tests, Including Spins, of a Free-Flying Radio-Controlled 0.13-Scale Model of a Twin-Jet Swept-Wing Fighter Airplane," NASA TM-SX-445 (1960).

5. Charles E. Libby and Sanger M. Burk, Jr., "Large-Angle Motion Tests, Including Spins, of a Free-Flying Dynamically Scaled Radio-Controlled 1/9-Scale Model of an Attack Airplane," NASA TM-X-551 (1961).

6. Henry A. Lee and Charles E. Libby, "Incipient- and Developed-Spin and Recovery Characteristics of a Modern High-Speed Fighter Design with Low Aspect Ratio As Determined from Dynamic-Model Tests," NASA TN-D-956 (1961).

7. Chambers, *Partners in Freedom*.

8. William P. Gilbert and Charles E. Libby, "Investigation of an Automatic Spin Prevention System for Fighter Airplanes," NASA TN-D-6670 (1972).

9. William P. Gilbert, Luat T. Nguyen, and Roger VanGunst, "Simulator Study of Applications of Automatic Departure- and Spin-Prevention Concepts to a Variable-Sweep Fighter Airplane," NASA TM-X-2928 (1973).

10. Holleman, "Summary of Flight Tests" NASA TN-D-8052.

11. Fratello, et al., "Use of the Updated NASA Langley Radio-Controlled Drop-Model Technique," AIAA Paper 87-2559; Croom, et al., "Dynamic Model Testing of the X-31," AIAA Paper 93-3674.

12. Croom, et al., "Research on the F/A-18E/F," AIAA Paper 2000-3913.

13. Bowman and Burk, "Stall/Spin Studies Relating to Light General-Aviation Aircraft," *Society of Automotive Engineers Business Aircraft Meeting, Wichita, KS*.

14. James S. Bowman, Jr., H. Paul Stough, III, Sanger M. Burk, Jr., and James S. Patton, Jr., "Correlation of Model and Airplane Spin Characteristics for a Low-Wing General Aviation Research Airplane," AIAA Paper 78-1477 (1978).

15. Staff of the Langley Research Center, "Exploratory Study of the Effects of Wing-Leading-Edge Modifications," NASA TP-1589 (1979).

16. Long P. Yip, Holly M. Ross, and David B. Robelen, "Model Flight Test of a Spin-Resistant Trainer Configuration," AIAA Paper 88-2146 (1988).

17. M.L. Holcomb and R.R. Tumlinson, "Evaluation of a Radio-Control Model for Spin Simulation," SAE Paper 77-0482 (1977); Holcomb, "The Beech Model 77 'Skipper' Spin Program," AIAA 1979-1835; R.R. Tumlinson, M.L. Holcomb, and V.D. Gregg, "Spin Research on a Twin-Engine Aircraft," AIAA Paper 1981-1667 (1981).

18. Burk and Wilson, "Radio-Controlled Model Design and Testing Techniques," NASA TM-80510.

19. Yip, et al., "Model Flight Test of a Spin-Resistant Trainer," AIAA 88-2146.

20. Cunningham, et al., "Practical Application of a Subscale Transport Aircraft for Flight Research," AIAA 2008-6200.

21. Robert W. Kamm and Philip W. Pepoon, "Spin-Tunnel Tests of a 1/57.33-Scale Model of the Northrop XB-35 Airplane," NACA Wartime Report L-739 (1944).

22. Ralph W. Stone, Jr., and Lee T. Daughtridge, Jr., "Free-Spinning, Longitudinal-Trim, and Tumbling Tests of 1/17.8-Scale Models of the Cornelius XFG-1Glider," NACA MR No. L5K21 (1946); Henry A. Lee, "Free-Spinning and Tumbling Characteristics of a 1/20-Scale Model of the Douglas XFG 4D-1 Airplane as Determined in the Langley 20-Foot Free-Spinning Tunnel" NACA RM-SL50K30a (1958); Lawrence J. Gale, Ira P. Jones, Jr., and Jack H. Wilson, "An Investigation of the Spin, Recovery, and Tumbling Characteristics of a 1/20-Scale Model of the Northrop X-4 Airplane," NACA RM-L9K28 (1950).

23. Raymond D. Whipple, Mark A. Croom, and Scott P. Fears, "Preliminary Results of Experimental and Analytical Investigations of the Tumbling Phenomenon for an Advanced Configuration," AIAA Paper 84-2108 (1984).

24. Charles E. Libby and Joseph L Johnson, Jr., "Stalling and Tumbling of a Radio-Controlled Parawing Airplane Model," NASA TN-D-2291 (1964).

25. G. Gratton and S. Newman, "The "Tumble" Departure Mode in Weightshift-Controlled Microlight Aircraft," Proceedings of the Institution of Mechanical Engineers, vol. 217, Part G: Journal of Aerospace Engineering (2003).

26. C.M. Fremaux, D.M. Vairo, and R.D. Whipple, "Effect of Geometry and Mass Distribution on Tumbling Characteristics of Flying Wings," NASA TM-111858 (1993).

27. W. Hewitt Phillips, Journey in Aeronautical Research-A Career at NASA Langley Research Center, NASA History Office Monographs in Aerospace History, No. 12 (Nov. 1998).

28. James S. Bowman, Jr., "Spin-Entry Characteristics of a Large Supersonic Bomber as Determined by Dynamic Model Tests," NASA TM-SX-1190 (1965).

29. Dennis Jenkins, Valkyrie: North American's Mach 3 Superbomber (Specialty Press, 2008).

Associated Test Techniques

1. Owens, et al., "Overview of Dynamic Test Techniques," AIAA Paper 2006-3146.

2. B.C. Trieu, T.R. Tyler, B.K. Stewart, J.K. Charnock, D.W. Fisher, E.H. Heim, J. Brandon, and S.B. Grafton, "Development of a Forced Oscillation System for Measuring Dynamic Derivatives of Fluidic Vehicles" 38th Aerospace Mechanisms Symposium, 2006, pp. 387–399.

3. J.M. Brandon, D.G. Murri, and L.T. Nguyen, "Experimental Study of the Effects of Forebody Geometry on High-Angle-of-Attack Static and Dynamic Stability and Control," Proceedings of 15th ICAS Congress, London, England, Sept. 1986, vol. 1 (New York: American Institute of Aeronautics and Astronautics, Inc., 1986), pp. 560–572.

4. Robert M. Hall, Shawn H. Woodson, and Joseph R. Chambers, "Accomplishments of the Abrupt Wing Stall (AWS) Program and Future Research Requirements," AIAA Paper 2003-0927 (2003).

5. Francis J. Capone, D. Bruce Owens, and Robert M. Hall, "Development of a Free-to-Roll Transonic Test Capability," AIAA Paper 2003-0749 (2003).

6. D. Bruce Owens, Jeffrey K. McConnell, Jay M. Brandon, and Robert M. Hall, "Transonic Free-To-Roll Analysis of the F/A-18E and F-35 Configurations," AIAA Paper 2004-5053 (2003).

7. James S. Bowman, Jr., Randy S. Hultberg, and Colin A. Martin, "Measurements of Pressures on the Tail and Aft Fuselage of an Airplane Model During Rotary Motions at Spin Attitudes," NASA TP-2939 (1989).

Future Perspectives

1. William L. Clarke and R.L. Maltby, "The Vertical Spinning Tunnel at the National Aeronautical Establishment, Bedford," R.A.E. Technical Note Aero 2339 (1954).

2. Robert M. Hall, C. Michael Fremaux, and Joseph R. Chambers, "Introduction to Computational Methods for Stability and Control (COMSAC)," *COMSAC Symposium, Hampton, VA, 2004.*

ABOUT THE AUTHOR

Joseph R. Chambers is an aviation consultant who lives in Yorktown, VA. He retired from the NASA Langley Research Center in 1998, after a 36-year career as a researcher and manager of military and civil aeronautics research activities. He began his career as a specialist in flight dynamics as a member of the staff of the Langley Full-Scale Tunnel, where he conducted research on a variety of aerospace vehicles, including V/STOL configurations, reentry vehicles, and fighter aircraft configurations. He later became the manager of research projects in the Full-Scale Tunnel, the 20-Foot Spin Tunnel, flight research at Langley, and piloted simulators. When he retired from NASA, he was manager of a group responsible for conducting systems analysis of the potential payoffs of advanced aircraft concepts and NASA research investments.

Mr. Chambers is the author of over 50 NASA technical reports and publications, including NASA Special Publications SP-514, on airflow condensation patterns for aircraft; SP-2000-4519, on contributions of Langley to U.S. military aircraft of the 1990s; SP-2003-4529, on contributions of Langley to U.S. civil aircraft of the 1990s; and SP-2005-4539, on Langley research on advanced concepts for aeronautics. He has made presentations on research and development programs to audiences as diverse as the von Karman Institute in Belgium and the annual Experimental Aircraft Association (EAA) Fly-In at Oshkosh, WI. He has served as a representative of the United States on international committees and has given lectures in Japan, China, Australia, the United Kingdom, Canada, Italy, France, Germany, and Sweden.

Mr. Chambers received several of NASA's highest awards, including the Exceptional Service Medal, the Outstanding Leadership Medal, and the Public Service Medal. He also received the Arthur Flemming Award in 1975 as one of the 10 Most Outstanding Civil Servants for his management of NASA stall/spin research for military and civil aircraft. He has a bachelor of science degree from the Georgia Institute of Technology and a master of science degree from the Virginia Polytechnic Institute and State University (Virginia Tech). He is coauthor of the recently released Radical Wings and Wind Tunnels by Specialty Press.

INDEX

Boeing 733 Supersonic Transport, 79

Boeing 757 aircraft, 40

Boeing 777 aircraft, 2

Boeing C-17 aircraft, 65

Boeing Phantom Works, 84, 87

Boeing wing stall research, 144

Boeing X-48A/X-48B Blended Wing-Body
  configuration, 5, 42, 83–84, 86

Boeing YC-14 aircraft, 65

Bomb release studies, 81, 82

Boundary layer flow, 14, 52

Bowman, Jim, 8

Brewster Buffalo airplane, 23

Brewster XF2A-1 aircraft, 47

British BAC 111 jet transport, 137–38

BT-9 aircraft, 47, 104, 105

Burk, Todd, 8

C

C-17 aircraft, 65

C-130 aircraft, 78

Calspan Transonic Wind Tunnel, 144

Cameras and video images
  catapult-launch techniques, 34
  drop-model testing, cameras on models during,
    35, 36, 37, 38, 120, 126, 131
  infrared cameras, 32
  powered free-flight models program, 41–42
  shadowgraph photography, 29, 31, 32, 53
  spin research, 19, 20, 120, 126, 131

Campbell, John P., 25, 65, 69

Canada, 22

Canards
  canard-type aircraft, 84
  dynamic stability and control, 50
  foldout for spin recovery, 112, 115, 126
  nose-down deflection, 3
  spin and spin recovery research, 112, 115
  spin tunnel tests, 105

Catapult-launch techniques, 18, 33–34, 120

Cayley, George, 1

Centrifugal forces (g-loads), 104, 126, 138

Century series of fighters, 42

Chance Vought F7U Cutlass, 86

Chance Vought F8U Crusader. See F8U Crusader

Chance Vought Low Speed Tunnel, 109

China, 22

Chined fuselage forebodies, 92, 143

Civil and commercial aviation programs. See also
  General-aviation aircraft; Supersonic Transport
  (SST)
    advanced civil aircraft, 83–84
    AirSTAR program, 39–40, 135–36
    fly-by-wire flight control systems, 80
    free-flight research, 44, 45
    free-to-roll tests, 143
    hypersonic research, 75
    model aircraft testing and, value of data from, 88
    NASA's relationship with, 39, 44
    powered free-flight models to study, 39
    results of tests, transmission of to, 44
    transport aircraft design and testing, 39–40, 64,
      83–84, 120, 135–36
    upset situations and recovery from, 39, 135–36
    VTOL aircraft, interest in, 58, 59
    wake vortex hazards, 84–85

Combat air patrol (CAP) missions, 69

Comet probe, 75

Composite materials, 34–35, 36

Compressed air ejectors, 27

Compressible flow, 10, 14, 42

Computational fluid dynamics (CFD), 148–49

Computational Methods for Stability and Control
  (COMSAC), 149

Computers, 1, 36, 38

Concept demonstrators, 86–88

Consolidated Vultee (Convair) XF-92 aircraft, 55

Consolidated Vultee (Convair) XFY-1 Pogo aircraft,
  59–60, 61

Constellation program, 75

Construction of models. See Design and
  construction of models

Control Line Facility, 40–41, 60

Control surfaces
  actuators to control, 13, 21, 22, 24, 25
  high angle of attack and, 99

Lockheed Martin F-22 fighter. *See* F-22 fighter

Lockheed Martin F-35 fighter, 144

Lockheed Martin wing-drop research, 144

Lockheed P-80A aircraft, 86

Lockheed Supersonic Transport, 79

Lockheed XFV-1 Salmon tail-sitter VTOL aircraft, 32, 59–61

Low-Speed Tunnel, 12-Foot. *See also* Free-Flight Tunnel, 12-Foot, 25, 82, 83, 96, 142, 145

Low-Turbulence Pressure Tunnel, 12-Foot, 29

Lunar Landing Facility, 138

M

M1 reentry concept, 77

M2-F1 lifting body, 38, 77, 78, 86

Mach numbers, 14, 82

Marine Corps, U.S., 94

Mars

    Aerial Regional-Scale Environmental Survey (ARES) of Mars, 72–73

    Mars airplane, 72–73

    space probes and planetary exploration, 74

Mars Science Laboratory program, 32

Mass (weight), 10, 11, 13

MCD-387-A fighter, 56

McDonnell-Douglas blended wing-body aircraft, 83

McDonnell-Douglas F-15 aircraft. *See* F-15 fighter

McDonnell-Douglas YC-15 aircraft, 65

McDonnell F3H Demon, 52

McDonnell F-4 Phantom II. *See* F-4 Phantom II fighter

McDonnell XF3H-1 Demon, 91

McDonnell XP-85 fighter, 57

Mercury program

    Pressurized Ballistic Range research for, 31

    space capsules, stability and control research on, 52, 73, 74

    Space Task Group for, 25

Microphone, 88

Military aircraft and programs. *See also specific fighter and attack aircraft*

    angle of attack research, 94–102

    blended wing-body designs, 84

Century series of, 42

chined fuselage forebodies, 143

development of, 94–102

development times, 147

drop-model testing, 4, 36, 37, 69, 119, 121–31

fly-by-wire flight control systems, 80

forced-oscillation tests, 142

free-flight model testing, 29, 39, 44, 45, 69, 94–102

free-to-roll tests, 142–43

fuselage shape and length, 50

hypersonic research, 75

Lightweight Fighter program, 96–97

losses, 94

model aircraft testing and, value of data from, 88

NASA's relationship with, 39

powered free-flight models to study, 39

rocket-boosted models and testing for, 52

scaling laws and procedures, 12

separation clearance studies, 80–83

spin and spin recovery research, 22

spin entry and poststall motion research, 119, 120–31

spin tunnel tests, 103, 105–15

transport aircraft design and testing, 64

variable-sweep configurations, 52, 68–69

Minesweepers, 86

Model 76 Duchess, 134

Model 77 Skipper Trainer, 117, 134

Model aircraft testing. *See also* Drop models and techniques; Free-flight models and tests, dynamic aerodynamic data, extraction of during, 139

    aircraft handling quality predictions and, 16, 147

    applications and roles for, 1–5

    correlation between full-scale flight tests and, 43, 47–48, 52, 90, 105, 108, 110–11, 120, 123, 128, 132–34

    correlation between theoretical predictions and, 50

    historical use of, 1, 16

    importance of use of, 1–2

    retention of models, 125

    success of tests, 16

T

T-28 aircraft, 115

T-34/YT-34 military trainer, 134

Tail-damping power factor (TDPF), 105, 106, 116

Tailless aircraft

  Kaiser Cargo Wing, 55

  spin tunnel tests, 105

  stability and control research, 49, 54–55, 87–88

  thrust vectoring and, 99, 100

  tumbling and, 136, 137

  X-36 Tailless Fighter Agility Research Aircraft program, 41–42, 86, 87–88

Tails

  all-movable tail concept, 75, 122

  differential tail, 75–76

  fins, 52

  flow separation, 91–92

  horizontal tails, 75, 93, 145

  rotary-balance testing and test apparatus, 145

  spin recovery and, 104, 105, 106, 116

  stability and control and, 49, 52

  T-tail, 145

  vertical tails, 9, 75, 91–92, 100, 104

  weather-vane stability, 91–92

Tail-sitter aircraft, 32, 41, 58–62, 114

Tailspin. *See also spin entries*, 18

Technical reports, 44

Thrust pilot, 26–27

Thrust vectoring, 98–100, 101

Thrust-vectoring paddles, 99–100

Thrust-vectoring vanes, 100

Tilted wind tunnel, 22

Tilt wing aircraft, 41, 60, 63–64

Time, 10

Tom-Tom, Project, 58

Towed vehicles research, 85–86

Trailing vortex wakes, 84–85

Transonic Dynamics Tunnel, 142, 144

Transonic flight

  funding for research programs, 71

  high-speed forced-oscillation tests, 142

  rocket-boosted models and testing, 42, 50–52

  stability and control and, 50–52

  unpowered research models, 42

  wind tunnel research, 42

Transonic Tunnel, 16-Foot, 62, 63, 144

Transport aircraft design and testing, 39–40, 64, 83–84, 120, 135–36

Tumbling, 22, 103–4, 136–37, 145

Turbofan engines, 13, 64–65

U

Unconventional aircraft development

  free-flight testing data, value of in, 7, 8–9

  free-flight testing of, 44

  research activities on, 147

  stability and control research, 53–58

Unitary Supersonic Wind Tunnel, 142

Unmanned aerial vehicles (UAVs), 147

Upper-surface-blowing (USB) concept, 64–65

V

V-173 Flying Pancake aircraft, 47, 54

Variable-sweep configurations

  folded aircraft concept, 72–73

  foldout wing concept, 71–72

  free-flight testing of, 68–73

  military aviation programs, 52, 68–69

  oblique wing (skewed wing) concept, 41, 69–71

  outboard pivot concept, 68–69

  rocket-boosted models and testing of, 52

  spin and spin recovery research, 107, 108, 112–13

  SST program, 79–80

  stability and control research, 68–73

Varieze aircraft, 84, 85

Vectored-thrust fighter program, 60, 62, 63

Vertical/short take-off and landing (V/STOL) aircraft. *See also* Short take-off and landing (STOL) aircraft, 41

Vertical tails, 9, 75, 91–92, 100, 104

Vertical take-off and landing (VTOL) aircraft

  advantages of, 58

  angle of attack research, 92

  aviation industry and government interest in, 58, 59

Wright, Wilbur, 1

X

Y

YF-107A aircraft, 75–76
YT-34 military trainer, 134

Z